# A $\mathscr{Motif}$ OF
# MATHEMATICS

## SCOTT B. GUTHERY

DOCENT PRESS
Boston, Massachusetts, USA
www.docentpress.com

Docent Press publishes monographs and translations in the history of mathematics for thoughtful reading by professionals, amateurs and the public.

The images on the cover and on page 96 (Figure 2.12), page 97 (Figure 2.13), and page 100 (Figures 2.14 and 2.15) are ©The Royal Society.

Cover design by Brenda Riddell, Graphic Details, Portsmouth, New Hampshire.

To Mary, my partner in all dances, plain and fancy.

Preface


In the introduction to his book, *History of Continued Fractions and Padé Approximants*, Claude Brezinski describes perfectly the path that I have followed.


> When I started this work, I soon realized that I was not a historian, that I knew nothing about the methods of history, and that it is in fact quite difficult to adequately approximate historian. [...] However, I hope that this history will be attractive to researchers working in the field, both from cultural and also from the mathematical point of view. Moreover, I would like it to be the starting point of some serious historical work. [**25**]


Let me quickly add that this by no means attempts to compare this modest monograph with Brezinski's magisterial study.

Almost every discussion of the Farey series in the mathematical literature makes fleeting reference to C. Haros as the person who first described the algorithm for creating this sequence of irreducible fractions. These references are often paired with dismissive remarks about John Farey by G. H. Hardy.

While Haros and Farey are walk-ons in the academic history of mathematics, fans of James Burke know there are fascinating stories in the hidden valleys between the peaks of greatness that resonate deeply with our every day lives.

So I decided to find out just who C. Haros was.

Charles Haros was a mathematician who worked with Lagrange, Legendre, Laplace, Lacroix, Delambre, Coulomb, de Prony, Lalande, Borda, Fourier, and Poisson at the turbulent confluence of the Bureau des Longitudes, the Ponts et Chaussées and École Polytechnique during an explosively creative period of French mathematics.

Haros was of a mathematical type of others we shall meet along the way. They are mathematicians whose output was quantitatively less than that of the greats but qualitatively on a par. They are also mathematicians who had the historical misfortune of living at a time when greats roamed the land.

Chuquet's régle des nombres moyens, Haros' sequence of irreducible fractions, Neville's tables, and Franel's Riemann equivalence are mathematical diamonds whose sparkle and radiance can be enjoyed for their own sake.

The primary intent of this monograph is to interest the reader in prospecting the newly-opened mines of digitized mathematical papers and books for gems of their own liking. If enjoyment happens to be found in reading about any of the ones I've found then so much the better.

Boston, Massachusetts
September 6, 2010

Acknowledgements

The first round of hearty huzzahs go to the heros at Google Books and the libraries around the world for their book scanning efforts. Free, on-line access to full text of thousands of rare and out-of-print books, papers and maps was a constantly flowing fountain of joy and surprise while researching and writing the monograph.

Special mention in this regard goes to Gallica, the website of the Bibliothéque nationale de France. Gallica established an early lead in providing free, open and unfettered access to the holdings of a national library. By continuing to add new content and advanced search capabilities the BnF has steadily increased that lead.

Thanks go to Paul Hamburg for translations from French and German and for a thorough editing pass. Mike Waterman and Gerard Kiernan found numerous potholes in the exposition that needed filling. Historians Hugh Torrens, David Grier, Alan Gluchoff, Tony Woolrich and Margaret Bradley provided much appreciated advice and guidance along the way.

Alex Yip did a deep dive in the Goodwyn archive at the Royal Society and came up with some pearls. In addition to imposing some order on the archive, his efforts help to bring this prodigious table maker into sharper focus. Thanks up top to Nichola Court, Felicity Henderson and Joanna Hopkins at the Royal Society for their help and hospitality during Alex's visits.

The friendly people at the Dolph Briscoe Center for American History in Austin helped me sort through the MTAC archives. The librarians at Harvard's Cabot Library helped me track down some of the originals that were scanned into Google Books.

Roger Mansuy over in Paris kindly shared his research on original papers by Haros. Keith Martin at NIST dug into dusty program listings from the days of the SEAC computer.

Credit and appreciation go to Brenda Riddell of Graphic Details for the stunning cover that captures the spirit of what's inside.

Mary Cronin challenged me to run a thread through scraps of reading, writing and computing about Farey sequence and the Riemann hypothesis that I'd accumulated over the years to see if they told a story. She was always there to help rethread the needle when diligence wavered.

There are undoubtedly errors, gaffs and howlers as there always will be when writing with an excess of enthusiasm and a paucity of discipline. I alone take responsibility for the route taken and the road signs missed.

The massive amount of book digitizing taking place is opening vast vistas of unexplored historical material. The contributions of amateurs to astronomy are acknowledged and appreciated by the professionals. Amateur historians of mathematics now have their own expanding universe to explore.

# Contents

# List of Figures

# List of Tables

# CHAPTER 1

# The Mediant

The following quotation from David Fowler's article, "An Approximation Technique, and its Use by Wallis and Taylor" is the inspiration for this monograph:

> This mediant has a long history going back to classical Greece: a special case of it is to be found at PLATO'S Parmenides 154bd, and, while the inequality does not occur in EUCLID'S Elements, the case of equality was treated as V 12 and VII 12. It was then stated and proved in PAPPUS, Collection VIII 8. CHUQUET rediscovered it and called it "la rigle des nombres moyens" (sic) in his Triparty en la Science des Nombres (1484).2 It was, we shall see, adduced by both WALLIS and, in a variant form, by TAYLOR. It was the first theorem enunciated in CAUCHY'S Cours d'Analyse (1821). It is also, inter alia, the basic generating property of FAREY series. Thus we have there an enduring *motif* of mathematics.[**81**]

The mediant as a *motif* of mathematics – that notion seemed to me reason enough to collect together in one place a more detailed history and discussion of a mathematical construct that is often merely mentioned in passing.

## 1.1. Non-Arithemetised Mathematics

We will discover that the mediant is more than just a simple computation that surfaces from time to time and in various mathematics contexts: the mediant is a different kind of computation than, say, the computation of a square root. What Fowler means by a being a motif is that the mediant embodies a particular way of thinking about mathematics.

Bringing the mediant to the fore lets us more clearly see the modes of thinking of the mathematicians who used it, and provides us with insight into their way of doing mathematics. Thinking of the mediant as a historical marker in mathematical studies is how Fowler came to the mediant. But by virtue of being a different and distinguished mode of mathematical thinking, the mediant yields unique mathematical insights and generates additional mathematical tools.

One of the pillar's of Fowler's analysis in his crowning work, "The Mathematics of Plato's Academy, A New Reconstruction" [**84**] is a fine-grain differentiation between ratio, proportion and fraction. This is a differentiation that is not made with any conviction or enthusiasm today today, but Fowler argues that this differentiation is critical to the understanding Greek mathematics, and for our purposes, one that yields differentiation and new mathematical tools.

Fowler argues that one must analyze arguments based on ratios and proportions in a different way than one analyzes arguments based on fractions. And he insists that analyzing arguments based on ratios and proportions as if they were based on fractions can lead to incomplete or even incorrect conclusions. In Fowler's view manipulating the ratio 2-of-3 is a different kind of mathematics – a different way of mathematical thinking – than manipulating the fraction two thirds.

Fowler refers to reasoning with fractions as *arithmetized mathematics*. "It is characterized by the use of some idea of number that is sufficiently general to describe some model of what is now frequently call 'the positive number line' *and its arithmetic.*" [**84**], p.8. Arithmetized mathematics builds constructs from numbers, and then reasons about these constructs using rules – arithmetics – for the manipulation of these numbers.

Non-arithmetized mathematics reasons about objects directly, using ratios and proportions without the initial step of modeling properties with numbers and converting manipulations on the object to manipulations of these numbers.

## 1.2. Ratio, Proportion and Fraction

While it is universally taken as the mark of a particularly lazy author to appeal to a dictionary for a definition, in the case of the three words in the title of this section this author is willing to don the red letter because there is widespread confusion today about these three words. Indeed in some circles – even mathematics circles – they are taken to be synonymous. In our considerations not only are they not synonymous but it is necessary that we are crystal clear about what we mean by each word. Thus, we slouch over to the Oxford English Dictionary.

DEFINITION 1. *ratio:* **2.a.** *Math.* The relationship between two similar magnitudes in respect of quantity, determined by the number of times one contains the other (integrally or functionally).

The first use of the word 'ratio' in the English language according to the O.E.D. is exactly the one we will adopt:

**1660** BARROW *Euclid* v. 3. Ratio (or rate) is the mutual habitude or respect of two magnitudes of the same kind each to other, according to quantity.

Following current practice we will write a ratio as $a : b$ and refer to $a$ as the first term of the ratio and $b$ as the second term of the ratio. Furthermore, for clarity of exposition we will always take the numerical value of $a$ to be less than or equal to the numeric value of $b$.

DEFINITION 2. *proportion:* **9.** *Math.* An equality of ratios, esp. of geometrical ratios; a relation among quantities of the first divided by the second is equal to that of the third divided by the fourth.

The uses that are closest in both time and meaning to our work are the following:

**1778** HUTTON *Course Math* I.110 If two or more couplets of numbers have equal ratios, or equal differences, the equality is named Proportion, and the terms of the ratios Proportionals.

> **1859** Barn. Smith *Arith. & Algebra* (ed. 6) 432 Proportion
> is the relation of equality subsisting between two ratios.

Thus, a proportion is the statement of the equality of two ratios. Two ratios are equal if and only if the first term of the first ratio divided by the second term of the first ratio is numerically equal to the first term of the second ratio divided by the second term of the second ratio.

Following current practice we will write a proportion as $a : b :: c : d$.

Definition 3. *fraction:* **5.** *Math.* **a.**\ *Arith.* A numerical quantity that is not an integer; one or more aliquot parts of a unit or whole number; an expression for a definite proportion of a unit or magnitude.

> **1827** Hutton *Course Math* I. 86 The vulgar fraction may be
> reduced to a decimal, then joined to the integer, and the root
> of the whole extracted.

A fraction denotes a specific part of a whole; half of a pie, a fifth of scotch, a quarter dollar and so forth. There are many ways of writing a fraction. For example, 0.5 of a pie, $\frac{1}{5}$ of a gallon of scotch, 20% of a dollar and so forth.

While a fraction is one number, a ratio is two numbers, and a proportion is four numbers, it is not difficult to see why they have become confused in our completely arithmetized world. A fraction is the number you get when you divide the two parts of a ratio in the process of determining if it is proportional to another ratio. In the arithmetized world, the model – the fractions – are the objects of study so our methodology instructs us to mute the reality from which they came – the ratios. For us the fractions are the ratios.

In Fowler's non-arithmetized world [**82**] there were no fractions. There were only ratios and statements of equality between them, proportions. Reasoning was done in reality and ratios provided a general-purpose way of reasoning that was independent of the units in which that reality was measured.

## 1.3. Definition of the Mediant

We are in position to define the mediant. We give both non-arithmetic and arithmetic definitions because both will show up in our work.

DEFINITION 4. The non-arithmetic mediant is the ratio $a + c : b + d$ computed from the two ratios $a : b$ and $c : d$.

If the ratios $a : b$ and $c : d$ are in proportion, $a : b :: c : d$, then the mediant satisfies the proportion too

$$a : b :: a + c : b + d :: c : d$$

DEFINITION 5. The arithmetic mediant is the fraction $\frac{a+c}{b+d}$ computed from the two fractions $\frac{a}{b}$ and $\frac{c}{d}$;

If the two fractions are such that $\frac{a}{b} < \frac{c}{d}$ then

$$\frac{a}{b} < \frac{a+c}{b+d} < \frac{c}{d} \qquad (1.3.1)$$

All would agree that the mediant is the proper method to combine two ratios and is the wrong way to combine two fractions if by combining fractions we mean to add them. If $a$ people in a population of $b$ people have brown hair and $c$ people in another population of $d$ people have brown hair, then $a + c$ people in the combined population of $b + d$ people have brown hair but $\frac{a+c}{b+d}$ is not the same as $\frac{a}{b} + \frac{c}{d}$.

In the following we will consider the use of the mediant in a number of mathematical settings and application contexts. As the usefulness of the mediant was recognized and harnessed over time, its definition has been generalized beyond 1.3.1. In some cases these generalizations are attuned to a specific problem or field of investigation. We will for the most part confine our attention to the domains of number theory and real analysis where mediant is of the general form

$$\frac{\alpha a + \beta c}{\alpha b + \beta d}$$

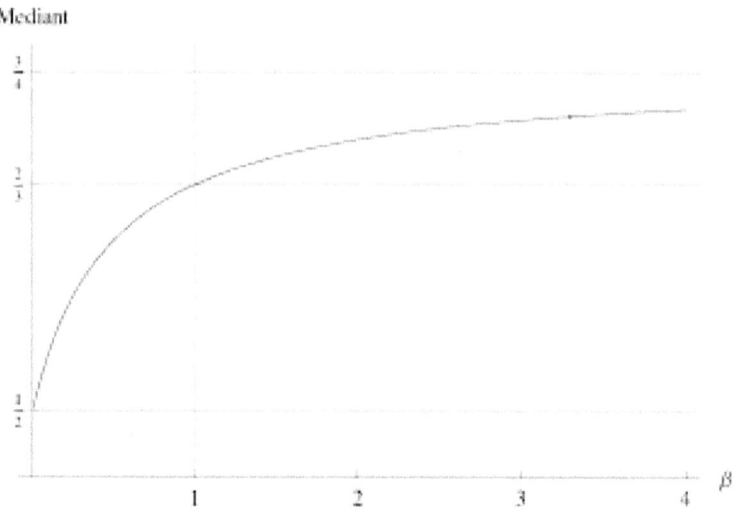

FIGURE 1.1. Mediants Between $\frac{1}{2}$ and $\frac{3}{4}$ for $\alpha = 1$

Figure 1.1 plots the mediants between $\frac{1}{2}$ and $\frac{3}{4}$ as a function of the mediant parameter $\beta$ when $\alpha = 1$. Note that mediant of Equation 1.3.1 where $\alpha$ and $\beta$ are both equal to 1 is heavily biased toward the larger of the two fractions.

Because of its association with the wrong way to add fractions, the mediant is often dismissed *in toto* or, what's worse, used as an example of ill-taught mathematics. V.I. Arnol'd took French mathematics education to serious task for being disconnected from reality. In the extended text of his address on teaching of mathematics given in the Palais de Découverte in Paris on March 7, 1997, he says,

> Rephrasing the famous words on the electron and atom, it can be said that a hypocycloid is as inexhaustible as an ideal in a polynomial ring. But teaching ideals to students who have never seen a hypocycloid is as ridiculous as teaching addition of fractions to children who have never cut (at least mentally) a cake or an apple into equal parts. No wonder that the children

will prefer to add a numerator to a numerator and a denominator to a denominator. [**7**]

## 1.4. A Sequence of Vulgar Fractions

The mediant computation of Equation 1.3.1 is most widely known though its association with the Farey fractions and as a result, the consideration of the Farey fractions is at the core of this monograph. As Fowler notes, however, the mediant has a much longer history than the Farey fractions and we will also discuss applications that are independent of the Farey fractions. Nevertheless, the Farey fractions are the natural place to start our history of the mediant.

The Farey fractions are the vulgar fractions – today we say would say irreducible fractions, reduced fractions or fractions in lowest terms – between 0 and 1 with denominators less than or equal to some given value. When these fractions are arranged in increasing order the result is the Farey sequence. For example, the Farey sequence when the given value is 5 is

$$\frac{0}{1}, \frac{1}{5}, \frac{1}{4}, \frac{1}{3}, \frac{2}{5}, \frac{1}{2}, \frac{3}{5}, \frac{2}{3}, \frac{3}{4}, \frac{4}{5}, \frac{1}{1}$$

The maximum denominator is called the *order* of the sequence and $F_m$ is used to denote the sequence of order $m$. While the Farey sequence is truly a sequence and not a series, the Farey sequence is also incorrectly called the Farey series. Depending on how the sequence is to be used, one or the other or both of the end-points, $\frac{0}{1}$ and $\frac{1}{1}$, may be excluded. Unless noted otherwise, we will always include both.

There are about $\frac{3}{\pi^2}m^2$ fractions in $F_m$. Figure 1.2 is a plot ratio of these two numbers. To be more precise, the number of fractions in $F_m$ is

$$\text{Length}(F_m) \approx \frac{3}{\pi^2}m^2 + O(m \ln n)$$

Figure 1.3 is the same ratio with the $m \ln m$ term added into the denominator.

You will immediately note, as did John Farey 195 years ago, that each fraction in this sequence of vulgar fractions is the mediant of its neighbors. This

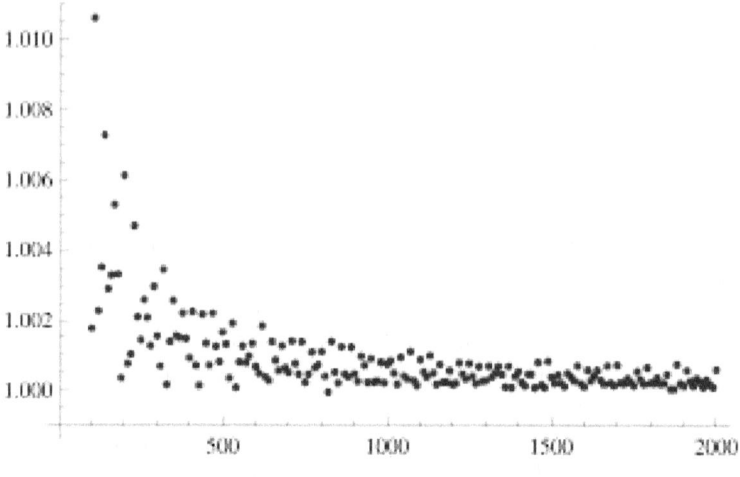

FIGURE 1.2. $\frac{\pi^2 \mathrm{Length}(F_m)}{3m^2}$ for $100(2000)10$

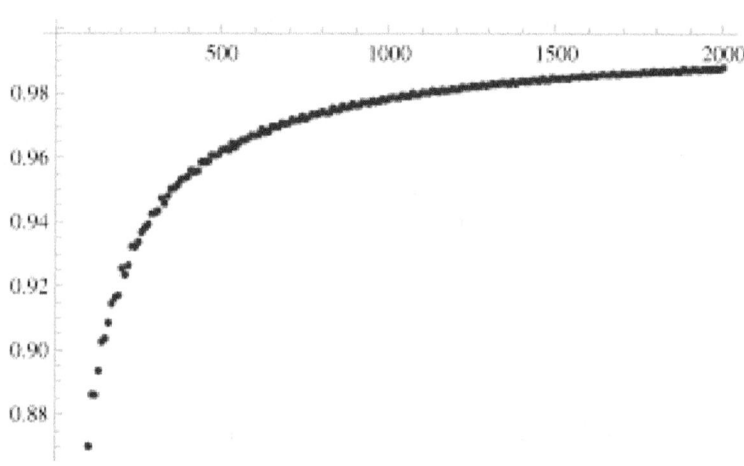

FIGURE 1.3. $\frac{\pi^2 \mathrm{Length}(F_m)}{3m^2 + m\ln m}$ for $100(10)2000$

property can be used to generate $F_{m+1}$ from $F_m$: starting with $F_m$ compute all the pairwise mediants, keeping those with denominator $m + 1$ and throwing away the others.

With the following Mathematica code

```
NextFarey[f0_] := Module[{f1 = {}, i, m},
  m = Denominator[f0[[2]]];
  For[i = 1, i < Length[f0], i++, f1 = Append[f1, f0[[i]]];
   If[Denominator[f0[[i]]] + Denominator[f0[[i + 1]]] == m + 1,
    f1 =
      Append[f1, (Numerator[f0[[i]]] +
          Numerator[f0[[i + 1]]])/(Denominator[f0[[i]]] +
          Denominator[f0[[i + 1]]])];
   ];
  ];
  Append[f1, f0[[i]]]
  ]
```

the Farey sequence of order m can be computed as

```
Nest[NextFarey, {{0, 1}, {1, 1}}, m-1]
```

Without recursion the Farey sequence of order m in Mathematica is simply

```
FareySequence[m_] :=
Join[{0},Union[Table[i/j,{j, m},{i,j-1}]//Flatten],{1}]
```

The mathematical literature abounds with papers describing and analyzing the properties of the Farey fractions and their numerous generalizations. An exhaustive summary of the current state of the analysis of Farey fraction is provided by Cristian Cobeli and Alexandru Zaharescu in their paper "The Haros-Farey Sequence at Two Hundred Years" [43].

There is no consensus as to when this sequence of vulgar fractions was first mentioned, but it was certainly over 500 years ago with some historians putting the figure at over 1500 years ago. Because the sequence is almost always discovered in a particular context or in the process of solving a particular problem, once discovered, the sequence has not had strong historical staying power in and of itself. As a result, it has been rediscovered time and again.

The simplicity of the Farey sequence together with the ease with which its unique properties can be stated and proven has meant that the fractions have been explored equally by amateur and professional mathematicians. That the sequence has attracted the attention of mathematicians of many levels of sophistication is not a property of the sequence but rather a property of the field of mathematics that it makes its home. Of all the diverse fields of mathematics the cooperation between amateurs and professionals is particularly characteristic of the theory of numbers. D.H. Lehmer [164] says

> Another peculiarity of the theory of numbers is the fact that many of its devotees are not professional mathematicians but amateurs with widely varying familiarity with the terminology and the symbolism of the subject.

Table 1.1 is tabular condensation the history of the Farey sequence from 1751 to 1914 as recounted by L.E. Dickson in *History and Theory of Numbers, Volume 1*, pp. 155-158. [62].

TABLE 1.1. Dickson's History of Farey Fractions from 1751 to 1914

| Year | Author | Contribution |
|------|--------|--------------|
| 1751 | Flitcon | Enumeration of irreducible fractions with denominators less than 100 |
| 1802 | Haros | Generation algorithm and demonstration of the mediant property |
| 1816 | Farey | Rediscovery of mediant the mediant property |
| 1816 | Cauchy | Generalized proof of the mediant property |
| 1818 | Goodwyn | Table of Farey fractions and observation of the mediant property |
| 1840 | Stouvenel | Proof of reflection around $\frac{1}{2}$ |
| 1850 | Eisenstein | Generalization and interpolation properties |
| 1858 | Stern | Generalization and interpolation properties |
| 1862 | Brocot | Extension beyond 1 |
| 1864 | Herzer | Tables of Farey fractions |
| 1876 | Hrabak | Tables of Farey fractions |
| 1876 | Halphen | Generalization of definition of Farey fractions |
| 1877 | Lucas | Generalization of definition of Farey fractions |
| 1879 | Glaisher | History of the Farey sequence |
| 1879 | Sang | Generalization of definition of Farey fractions |
| 1880 | Minie | Generalization of the enumeration of Farey fractions |
| 1881 | Pullich | New proof of the mediant property |
| 1881 | Airy | Tables of Farey fractions |
| 1883 | Sylvester | Enumeration of Farey fractions |
| 1886 | Ocagne | Generalization and extension beyond 1 |
| 1894 | Hurwitz | Application to rational approximation |
| 1894 | Hermes | Recursive definition of Farey numbers |
| 1895 | Vahlen | Application to the composition of linear fractional substitutions |
| 1903 | Made | Application to complex numbers |
| 1905 | Busche | Geometric generalization |
| 1909 | Sierpinski | Application to proof of irrationality |
| 1914 | Anonymous | Rediscovery of Farey fractions |

TABLE 1.2. Early History of a Sequence of Vulgar Fractions

| Year | Author | Reference |
|------|--------|-----------|
| 1685 | John Wallis | David H. Fowler [81] |
| 1563 | Juan Perez de Moya | Jacques Sesiano[217] |
| 1559 | Johannes Buteo | Jacques Sesiano[217] |
| 1520 | de la Roche | Graham Flegg [75] |
| 1512 | Juan de Ortega | Jacques Sesiano [217] |
| 1484 | Chuquet | Graham Flegg [75] |
| 1430 | Unknown | Jacques Sesiano [217] |
| 340 | Pappus | David H. Fowler [81] |
| 100 | Nicomachus | Jay Kappraff [145] |
| 250 BCE | Archimedes | Edouard Lucas [167] |
| 300 BCE | Euclid | David Fowler [84] |
| 325 BCE | Plato | David H. Fowler [81] |

Table 1.2 is a retrospective condensation of the history of the Farey sequence going back from 1750. The entries in the *Citation* column are exemplars and should not be interpreted as being either the first or only citation for the author's work with the Farey fractions.

Lucas [167] and Sesiano [217] both report that the mediant was known to Indian mathematicians, and in the same citation, Sesiano writes that it was known to Islamic mathematicians, without providing any more specific information.

From time to time, mathematicians working with the mediant were aware of the contribution of those going before but more often than not they weren't.

Of course, there may be a variety of reasons that prior work might have been overlooked but as we noted above, the mediant and the vulgar fractions it computes were part of a solution to a particular problem, and not the object of study per se.

In a posting to `math-history-list` in 1997, Fowler cites Plato writing in Parmenides 154b-d as follows:

> If one thing is older than another, it cannot be becoming older still, nor the younger younger still, by any more than their original difference in age ... But if an equal time is added to a greater time and to a less, the greater will exceed the less by a smaller part.

Fowler says goes on to say

> In other words(?) if $p : p < r : s$ (ie $r > s$) then $p : p < (p+r) : (p+s) < r : s$.
> In order, then, to bestow a (specious?) parentage on the mediant property, I call it the 'Parmenides Proposition'!

Jacques Sesiano [**217**] finds the mediant property in a Provençal manuscript written by an unknown author around 1430, which can be interpreted as suggesting that the mediant property be called the 'Pamiers algorithm', after the name of the Languedocian city in which the manuscript was written. If we were forced to commit ourselves to the moment of the first application of the mediant in the mathematical literature, we would agree with the general consensus that it would be in the work of Chuquet [**40**] where Chuquet dubs it the "règle des nombres moyens" and uses it an algorithm to compute roots. Cajori says [**31**]

> The earliest known process in the Occident of approaching to a root of an affected numerical equation was invented by Nicolas Chuquet, who in 1484 at Lyons, wrote a work of high rank entitled *Le triparty en la science des nombres*. It was not printed until 1880. If $\frac{a}{c} < x < \frac{b}{d}$, then Chuquet takes the intermediate value $\frac{a+b}{c+d}$ as a closer approximation to the root $x$.

The debate about the historically correct name for the sequence of vulgar fractions created by the mediant persists to this day, and mathematicians of the caliber of G. H. Hardy have weighed in on the topic. In the following we will refer to the sequence as the "Farey sequence," simply because this has become its common and familiar designation, and the name by which it is primarily known in the literature and, thereby, in the land of search engines.

## 1.5. Nicolas Chuquet and the Règle des Nombres Moyens

Nicolas Chuquet wrote a tract entitled *Le Triparty en la science des nombres* some time between 1480 and 1484 ([**40**], [**41**]). As far as is known there is only one copy of the manuscript extant. It is no. 1346 of the *Fonds français* in the Bibliothèque nationale de France.

There is a fascinating story of how the manuscript was lost and how it was found. This story together with a thorough exegesis of *Triparty*, including selected translations may be found in found in a book written by Graham Flegg, Cynthia Hay, and Barbara Moss, *Nicolas Chuquet, Renaissance Mathematician*[**75**].

In his description of Chuquet in the *Biographical Dictionary of Mathematicians* Jean Itard writes [**139**]:

> Chuquet made few claims of priority. The only thing that he prided himself on was his personal discovery was his "règle des nombres moyens."

While there are debates about earlier sightings of the mediant, there is no debate that Chuquet's règle des nombres moyens is the mediant and so 1484 is an upper bound for its first appearance. Chuquet's rule appears at the end of the first part of the *Triparty* manuscript. The paragraphs which describe the rule, as translated by Flegg et. al., are as follows:

> The rule of intermediate numbers
> 
> This rule serves to find as many numbers intermediate between two neighbouring numbers as one desires. By its means it is possible to find many more numbers and to do more calculations than are found by the rule of three or by one position or by two positions. And to understand and know how to practise this rule, one should know that $\frac{1}{2}$ is the first and beginning among the fractions, and from it arise and issue forth two natural progressions, of which one proceeds by increasing, as $\frac{1}{2}$,

$\frac{2}{3}$, $\frac{3}{4}$, $\frac{4}{5}$, etc. and the other proceeds by diminishing, as $\frac{1}{2}$, $\frac{1}{3}$, $\frac{1}{4}$, $\frac{1}{5}$, etc. These things being understood, the rule follows,

Numerator is added to numerator, and denominator to denominator. This means that when one wishes to find the first intermediates between two neighbouring whole numbers, one should add $\frac{1}{2}$ to the smaller whole number, and thus one will have an intermediate number, greater than the smaller extreme and smaller than the greater extreme.

... And whoever would want to find an intermediate between $3\frac{1}{3}$ and $3\frac{2}{5}$ should manipulate according to the rule, and one will have $3\frac{5}{8}$. And in this way one may continue to search for intermediate numbers, until one has found what one was looking for.

The purpose for which Chuquet brings his rule into being is to solve equations of one variable and he uses it this for this purpose in the second and third parts of *Triparty*.

He does not demonstrate any specific properties of his rule of intermediate numbers, for example that the fractional parts are irreducible or that they satisfy the modular property, but in reading the first sentence, it appears that he does grasp a very important property; viz. that the rule of intermediate values fills in all the rational values between the initial extremes.

Some analysts of the *Triparty* have found precursors of the logarithm but this view is not widely held. Graham Flegg, a leading biographer of Chuquet, says ([**74**], p.140)

(Chuquet) may have been aware of the principle underlying logarithms, though he did not develop a detailed discussion of it.

This is not as unlikely a connection as it might appear at first blush. We will find logarithms and the Farey fractions intertwined in a subsequent chapter.

## 1.6. Rational Approximation

The *règle des nombres moyens* is the *régle de double fausse position* or, as it is more widely known, *regula falsi* for fractions. The contemporary use of this precursor to the *règle des nombres moyens* has recently been discussed at length in papers by Spiesser [**221**] and Lamassé [**156**].

Chuquet and many that followed him used the règle des nombres moyens as way to generate ever better approximate solutions in extracting roots and solving equations. The staring point is a pair of fractions that bracket the answer; $\frac{a}{b}$ being smaller than the answer and $\frac{c}{d}$ being larger than the answer. Chuquet computes the mediant of the brackets,

$$\frac{a}{b} < \frac{a+b}{c+d} < \frac{c}{d},$$

and then plugs the result into the problem statement to get a sense of how much closer he might be to a solution. If the mediant value is a solution or is sufficiently close to a solution, then he's done. If he wants a better solution, then the mediant replaces one of the brackets and he goes around again.

Table 1.3 displays Chuquet's mediant algorithm finding $\sqrt{2}$ and Figure 1.4 is a log plot of the absolute error. After 17 iterations we have the answer to 6 decimal places.

On purely numerical grounds there is nothing particularly exceptional about the mediant. We could just as well have used the arithmetic, geometric or harmonic mean of the brackets in the algorithm and found $\sqrt{2}$ to 6 decimal places in about the same number of iterations. The mediant offers something that was important to *algoristes* – people who computed for a living – such as Chuquet: speed. If you are working with pen and ink and have lots of iterations to perform to solve your problem, then any technique that reduces the cycle time of each iteration is bound to attract attention. Algorithmic cycle time was just as important in the $15^{th}$ century as it is in the $21^{st}$.

For rational approximation, Chuquet's mediant gradually gives way to the continued fraction expansions of Cataldi and Wallis, and Euler and Lambert. For solving equations it will give way to the *reguli-falsi* of Newton and increasingly robust and clever methods of solvers such as Lagrange, Laguerre,

TABLE 1.3. Règle des Nombres Moyens Seeks $\sqrt{2}$

|    | Low | Mediant | High | Mediant² | Error |
|----|-----|---------|------|----------|-------|
| 1  | 1 | 3/2 | 2 | 2.250000 | 0.250000 |
| 2  | 1 | 4/3 | 3/2 | 1.777777 | -0.222222 |
| 3  | 4/3 | 7/5 | 3/2 | 1.960000 | -0.040000 |
| 4  | 7/5 | 10/7 | 3/2 | 2.040816 | 0.040816 |
| 5  | 7/5 | 17/12 | 10/7 | 2.006944 | 0.006944 |
| 6  | 7/5 | 24/17 | 17/12 | 1.993079 | -0.006920 |
| 7  | 24/17 | 41/29 | 17/12 | 1.998810 | -0.001189 |
| 8  | 41/29 | 58/41 | 17/12 | 2.001189 | 0.001189 |
| 9  | 41/29 | 99/70 | 58/41 | 2.000204 | 0.000204 |
| 10 | 41/29 | 140/99 | 99/70 | 1.999795 | -0.000204 |
| 11 | 140/99 | 239/169 | 99/70 | 1.999964 | -0.000035 |
| 12 | 239/169 | 338/239 | 99/70 | 2.000035 | 0.000035 |
| 13 | 239/169 | 577/408 | 338/239 | 2.000006 | 0.000006 |
| 14 | 239/169 | 816/577 | 577/408 | 1.999994 | 0.000006 |
| 15 | 816/577 | 1393/985 | 577/408 | 1.999998 | 0.000002 |
| 16 | 1393/985 | 1970/1393 | 577/408 | 2.000001 | 0.000001 |
| 17 | 1393/985 | 3363/2378 | 1970/1393 | 2.000000 | 0.000000 |

and Halley, down to the present day solvers of Householder, and Berlekamp, to name just a few. These generalizations and elaborations avoid numerical potholes that would surely break the mediant's axle.

And yet the very speed with which the mediant could be computed in the days of hand computation also recommends it for time-constrained algorithms for today's high-speed computers. The success of chess playing programs from Belle to Deep Fritz is due in part to the balance they strike between a small number of sophisticated computations against a large number of naive computations. In the era of ubiquitous computers there is a second benefit to being simple and fast: software surety. There are limitations to every algorithm however general-purpose. The boundary between the domain where an algorithm works and where it doesn't is easier to find, easier to characterize and easier to avoid for a simple algorithm than for a complex algorithm. While generality is a much admired feature of mathematics, gratuitous generality – generality that

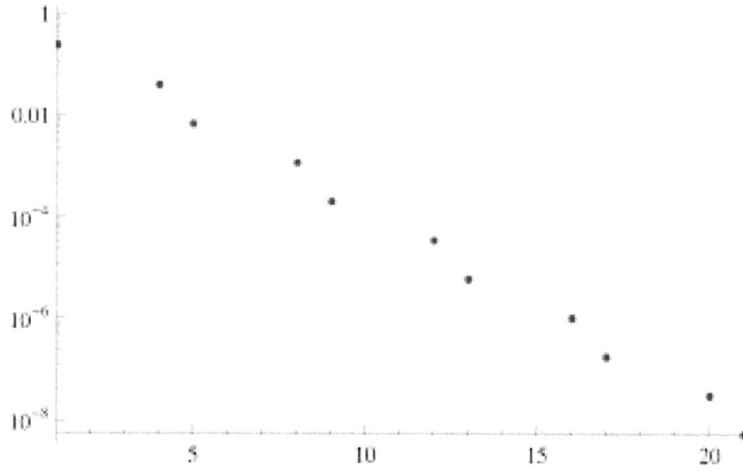

FIGURE 1.4.  Absolute Error of the Règle des Nombres Moyens
as it Seeks $\sqrt{2}$

is unneeded and unused – should evoke scepticism rather than unquestioned ad-
miration when it is found in software. If the problem is understood well, then
an optimized and tuned application of a simple algorithm such as Chuquet's
mediant may be sufficient both in terms of speed and safety.

```
RegleDesNombresMoyens[l_, h_, f_, t_, m_, e_]  :=
  Module[{i, low = l, mediant, high = h},
  For[i = 1, i < m, i++,
   mediant = (Numerator[low]   + Numerator[high])/
             (Denominator[low] + Denominator[high]);
   Which[
    Abs[f[mediant] - t] < 10^(-e), Break[],
    f[mediant] < t, low = mediant,
    True, high = mediant
    ];
   ];
  {i,mediant,f[mediant],N[f[mediant],6],N[f[mediant]-t,6]}
  ]
```

## 1.7. The Mediant and the Continued Fraction

Chuquet used his règle des nombres moyens – the mediant – as a quick and easy way to generate increasingly accurate trial values in extracting roots and solving equations. Starting with one fraction below the answer and another above the answer he proceeded as follows:

(1) Compute the mediant of the two fractions.
(2) IF the mediant solves the equation, THEN STOP
(3) IF the mediant is too small, THEN it becomes the lower fraction.
(4) IF the mediant is too large, THEN it becomes the larger fraction.
(5) GOTO 1.

Whether the trial value is too large or too small is determined by plugging it into the equation that is being solved. If making the trial value smaller would bring the equation closer to a solution, then the trial value is too large. If making the trial value a little larger would get the equation closer to a solution, then it is too small.

For example, if you are computing the square root of a number – as Chuquet was doing – you would square the trial value and compare the result to the number whose square root you are trying to compute. If the square of the trial value is larger than the number whose square root you're looking for, then the square root is somewhere between the trial value and the lower fraction. Set the upper fraction to the trial value and try again.

Now, suppose that each time we go around this loop, we tally up how many times in succession the trial value was too large or too small. For example, we might start off by seeing a run of $n_1$ values that too large, then a run of $n_2$ values that are too small, then $n_3$ values that are too large, and so forth.

For clarity of discussion, we'll assume that solution to the equation is between 0 and 1 and that $\frac{0}{1}$ and $\frac{1}{1}$ are the starting low and high fractions respectively. If this isn't the case, we begin by computing the integer part of the solution on the back of an envelope, and then go on to solve the equation for the fractional part.

The starting condition for the counting algorithm is a run of length 1 of too-large since the solution is less than 1 and so the first and current count, $n_1$, is equal to 1. Letting $n_2$ be the length of the following too-small run, $n_3$ the length of the next too-large run, and so forth, the lower and upper fractions at the end of each run are as follows:

$$\left\{ 0, \frac{1}{n_1} \right\}$$

$$\left\{ \frac{n_2}{1 + n_1 n_2}, \frac{1}{n_1} \right\}$$

$$\left\{ \frac{n_2}{1 + n_1 n_2}, \frac{1 + n_2 n_3}{n_1 + (1 + n_1 n_2) n_3} \right\}$$

$$\left\{ \frac{n_2 + (1 + n_2 n_3) n_4}{1 + n_1 n_2 + (n_1 + (1 + n_1 n_2) n_3) n_4}, \frac{1 + n_2 n_3}{n_1 + (1 + n_1 n_2) n_3} \right\}$$

$$\left\{ \frac{n_2 + (1 + n_2 n_3) n_4}{1 + n_1 n_2 + (n_1 + (1 + n_1 n_2) n_3) n_4}, \right.$$
$$\left. \frac{1 + n_2 n_3 + (n_2 + (1 + n_2 n_3) n_4) n_5}{n_1 + (1 + n_1 n_2) n_3 + (1 + n_1 n_2 + (n_1 + (1 + n_1 n_2) n_3) n_4) n_5} \right\}$$

Each of these fractions can be unwound into a form that is quite familiar today. Take, for example, the high fraction in the third pair.

$$\frac{1 + n_2 n_3}{n_1 + (1 + n_1 n_2)n_3} = \frac{1}{\dfrac{n_1(1 + n_2 n_3) + n_3}{1 + n_2 n_3}}$$

$$= \frac{1}{n_1 + \dfrac{n_3}{1 + n_2 n_3}}$$

$$= \frac{1}{n_1 + \dfrac{1}{\dfrac{1 + n_2 n_3}{n_3}}}$$

$$= \frac{1}{n_1 + \dfrac{1}{n_2 + \dfrac{1}{n_3}}}$$

If we run the mediant counting algorithm with the target being $\sqrt{\pi} - 1$ the first five counts are

$$\{n_1, n_2, n_3, n_4, n_5\} = \{1, 3, 2, 1, 1\}$$

Plugging these counts into 1.7.1 we get

$$\left\{\frac{10}{13}, \frac{17}{22}\right\} \approx \{0.769231, 0.772727\}$$

```
MediantApproximation[x_, n_, lower_, upper_] :=
 Module[{i, mediant, pl, ql, pu, qu,
         count = 1, digits = {0}, over},
  pl = Numerator[lower]; ql = Denominator[lower];
  pu = Numerator[upper]; qu = Denominator[upper];
  mediant = (pl + pu)/(ql + qu);
  over = mediant < x;
  For[i = 1, i <= n, i++,
   mediant = (pl + pu)/(ql + qu);
   If[(mediant < x && over) || ( mediant > x && ! over),
    digits = Append[digits, count];
   over = ! over;
```

```
count = 0,
If[mediant < x ,
 pl = Numerator[mediant]; ql = Denominator[mediant],
 pu = Numerator[mediant]; qu = Denominator[mediant];
 ];
count++;
];
];
digits
]
```

David Fowler says in [83] "The first explicit hint of continued fractions is found in Rafael Bombelli's *Algebra* published in Bologna in 1572. ...we find very similar expressions in Pietro Cataldi's *Trattato del modo brevissimo di trovar la Radice quadra delli numeri* ... (1613, also published in Bologna but without any reference to Bombelli); ...". Claude Brezinski's book *History of Continued Fractions and Padé Approximants* [25] is an exhaustive study and could justifiably be regarded as the definitive work about its title subject.

Brezinski calls Peitro Antonio Cataldi "the real discoverer of the theory of continued fractions" on the basis of Cataldi's 1613 tract. Brezinski also credits Cataldi with the now-familiar notation of continued fractions. Cataldi's focus was on computing square roots. It is not crystal clear from Fowler's or Brezinski's description of Cataldi's work that Cataldi was using the mediant to generate trial values for his square root computations. What is certainly true is that in one way or another, Cataldi came up with the digits of the simple continued fraction expansion; i.e. the run counts.

### 1.8. John Wallis, Savilian Chair of Geometry

Regarding Wallis, Brezinski says, "After the first attempts of Bombelli and Cataldi, the theory of continued fractions was ready to be developed. The beginning of the theory was due mostly to Wallis ..." How beginning the theory differs from discovering the theory is not clear. David Fowler [81] did a careful reading of Wallis' 1685 tract *A Treatise of Algebra, both Historical and Practical, Shewing the Original, Progress, and Advancement thereof, from time to time, and by what Steps it hath attained to the Height at which it now is.* Wallis used a number of different methods to compute the digits of the continued fraction one of which was the mediant as described above. Another was the Euclidian algorithm.

(1)  $1 = 1 \times (1 - \sqrt{\pi}) + (2 - \sqrt{\pi}) = 1 \times 0.772454 + 0.227546$
(2)  $0.772454 = 3 \times 0.227546 + 0.0898154$
(3)  $0.227546 = 2 \times 0.0898154 + 0.0479153$
(4)  $0.0898154 = 1 \times 0.0479153 + 0.0419001$
(5)  $0.0479153 = 1 \times 0.0419001 + 0.00601528$

Wallis used the equality of the expansion digits, $\{1, 3, 2, 1, 1, \ldots\}$, as proof of the equivalence of the two methods and then imputed properties he discovered in one to the other. While this proof falls short of today's expectations for rigor, Wallis demonstrated a direct connection between the mediant and the Euclidean algorithm and in the process, provided the first indication that there is something more about the mediant function than just a quick way of generating increasingly interesting trial values.

## 1.9. Digit Generation

Once the wider utility of the continued fraction representation was perceived, its form quickly suggested yet a third way to compute the digits of the expansion. Not only is this method easier, but it also facilitates the use of the continued fraction representation of numbers in theoretical investigations. One starts with a transformation called the Gaussian map

$$T(x) = \left\langle \frac{1}{x} \right\rangle$$

and finds

$$n_i(x) = \left[ \frac{1}{T^{i-1}(x)} \right]$$

The numbers circulating through these three continued fraction expansion algorithms are all different and thus one might be used in preference to another in a particular situation.

TABLE 1.4. Numbers Circulating in Digit Algorithms

| Digit | Parmenides Algorithm | Euclidean Algorithm | Gaussian Map |
|---|---|---|---|
| 1 | $\left(\frac{0}{1}, \frac{1}{1}\right)$ | $(0.772454, 0.227546)$ | $0.294576$ |
| 3 | $\left(\frac{3}{4}, \frac{1}{1}\right)$ | $(0.227546, 0.0898154)$ | $0.394713$ |
| 2 | $\left(\frac{3}{4}, \frac{7}{9}\right)$ | $(0.0898154, 0.0479153)$ | $0.533487$ |
| 1 | $\left(\frac{10}{13}, \frac{7}{9}\right)$ | $(0.0479153, 0.0419001)$ | $0.87446$ |
| 1 | $\left(\frac{10}{13}, \frac{17}{22}\right)$ | $(0.0419001, 0.00601528)$ | $0.143563$ |

Readers interested in an update on the state of rational approximation algorithms might consider including "A Comparison of Algorithms for Rational $l_\infty$ Approximation" by C.M. Lee and F.D.K. Roberts [162] on their reading list.

## 1.10. The Möbius Transformation

Laubenbacher, McGrath and Pengelley in [159] describe Lagrange's algorithm for finding the roots of equations using continued fractions [153]. At each step of the algorithm Lagrange transforms the equation being solved by substituting $n + \frac{1}{x}$ for $x$ where a lone root is known to be contained in the integer interval, $(n, n+1)$. Writing this transformation of variable as

$$\chi_n(x) = \frac{nx + 1}{x + 0} \tag{1.10.1}$$

sequential repetitions of this algorithmic step transform the original equation $f$ into the $f$ evaluated at the composition of these transformations,

$$f(\chi_{n_1}(\chi_{n_2}(\dots x \dots))),$$

and furthermore the convergents can be computed using matrix multiplication.

In particular, if we set

$$N_i = \begin{pmatrix} n_i & 1 \\ 1 & 0 \end{pmatrix}$$

then

$$\begin{pmatrix} q_n & q_{n_1} \\ p_n & p_{n-1} \end{pmatrix} = \prod_{i=1}^{n} N_i$$

where

$$\frac{p_n}{q_n}$$

is the $n^{th}$ convergent.

The name associated with transformation 1.10.1 is Möbius. The transformation itself is called a linear fractional transformation, a homographic transformation or simply a Möbius transformation. Möbius [181] re-discovered the transformation in a geometric setting rather than in an equation-solving setting but the transformation served exactly the same purpose, to produce an analytically useful mediant value between two existing values. In Möbius' case it was to produce a point between two given points or a line between two given lines rather than a rational number between two given rational numbers.

Wallis, Lagrange and the equation solvers composed a series of transformations, each of which represented a more accurate approximation to a solution. Rendering transformation composition as matrix multiplication made it easier to apply the continued fraction algorithm and led to a generalization – also described by Laubenbacher, et. al. – that improved the speed of convergence rate of the algorithm.

Möbius, Chasles and the geometers composed transformations each of which was a geometric primitive, such as a rotation or a translation of the geometric object. Rendering transformation composition as matrix multiplication provided them with a representation of the transformation that was the combination of all of the primitives.

## 1.11. Mediant Convergents

Using Möbius transformations Shunji Ito has described two additional convergent sequences [140]. The one of interest to us is the following. Let

$$T(x) = \begin{cases} x/(1-x), \ 0 \le x < 1/2 \\ (x-1)/x, \ 1/2 \le x < 1 \end{cases} \tag{1.11.1}$$

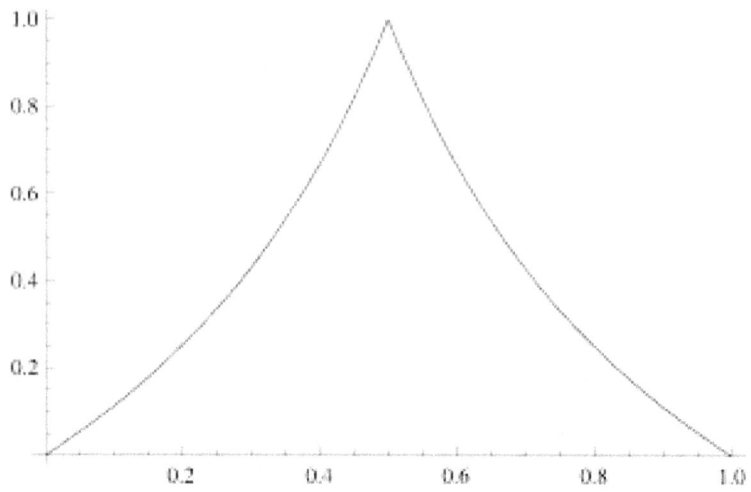

FIGURE 1.5. Mediant Transform Function

and set

$$\begin{pmatrix} r_n & s_n \\ t_n & u_n \end{pmatrix} = \prod_1^n A_{\epsilon_k}(x)$$

where

$$\epsilon_k(x) = I^{[1/2,1]}(T^{n-1}(x)).$$

The fractions in the resulting sequence

$$\frac{p_n}{q_n} = \frac{t_n(x) + u_n(x)}{r_n(x) + s_n(s)}$$

are the mediant convergents to $x$.

Figure 1.5 is the transform $T(x)$ and Figure 1.6 is a plot of Ito's function for $n = 5$. The vertical lines are drawn at Farey values of order 8. It's easy to see why 1.11.1 is called the Farey map. What some may find harder to understand is how prime factorization is encoded in such a simple transformation.

$$\{\frac{0}{1}, \frac{1}{8}, \frac{1}{7}, \frac{1}{6}, \frac{1}{5}, \frac{1}{4}, \frac{2}{7}, \frac{1}{3}, \frac{3}{8}, \frac{2}{5}, \frac{3}{7}, \frac{1}{2}, \frac{4}{7}, \frac{3}{5}, \frac{5}{8}, \frac{2}{3}, \frac{5}{7}, \frac{3}{4}, \frac{4}{5}, \frac{5}{6}, \frac{6}{7}, \frac{7}{8}, \frac{1}{1}\}.$$

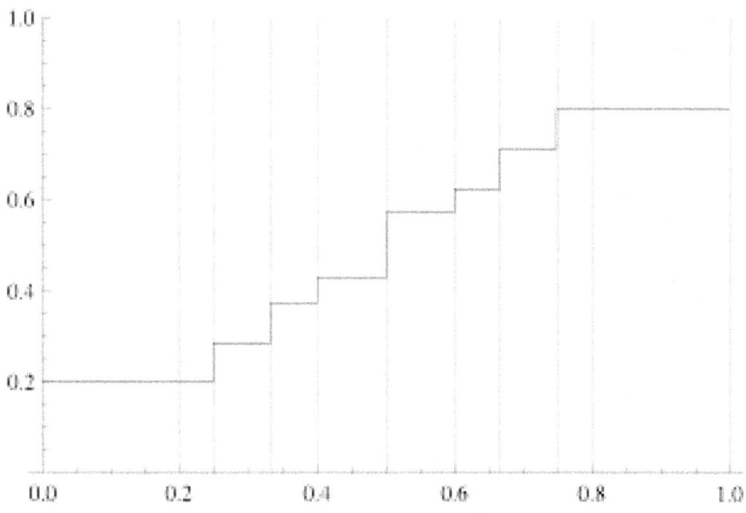

FIGURE 1.6. $p_4(x)/q_4(x)$ with grid at $F_5$

The sequence of mediant convergents is exactly the same as the sequence of fractions generated by the successive use of the mediant in the Parmenides Algorithm. Fowler gives a beautiful geometric interpretation of the sequence and its generation in [**84**].

```
Epsilon[x_] := Which[x >= 1/2, 1, True, 0]

T[x_] := (1 - \[Epsilon][x]) x /(1 - x) + Epsilon[x] (1 - x)/x

MediantConvergent[x_, n_] :=
 Module[{i, a = IdentityMatrix[2], t = x},
  For[i = 1, i <= n, i++,
   a = a.{{1, 1}, {Epsilon[t], 1 - Epsilon[t]}};
   t = T[t];
   ];
  {a[[1, 1]], a[[1, 2]], a[[2, 1]], a[[2, 2]]}
  ]
```

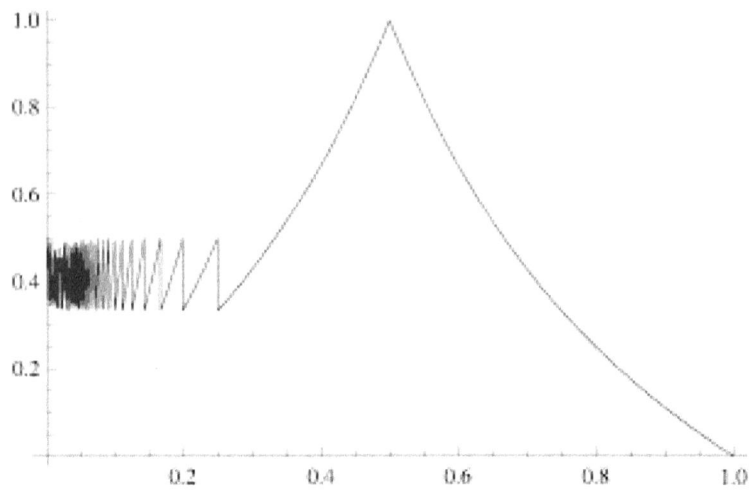

FIGURE 1.7. Nearest Mediant Transform Function

In the same paper Ito also succeeded in building a transform function for the nearest mediant continued fraction expansion. It is plotted in Figure 1.7 and its convergents are plotted in Figure 1.8.

```
J[x_] := If[FractionalPart[1/x] == 0,
          IntegerPart[1/x] - 1,  IntegerPart[1/x]]

T1[x_] := Module[{k}, k = J[x];
  Which[k == 1, (1 - x)/x, k == 2, x/(1 - x), True,
   x/(1 - (k - 2) x)]
 ]

B[k_] := Module[{i, b = IdentityMatrix[2]},
  Which[
   k == 1, Return[{{1, 1}, {1, 0}}],
   k == 2, Return[{{1, 1}, {0, 1}}],
   True,
   For[i = 3, i <= k, i++,
    b = b.{{1, 1}, {0, 1}}
```

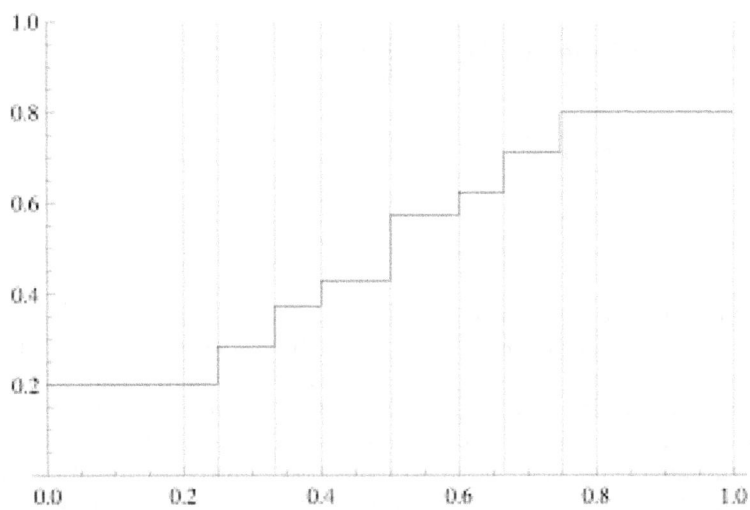

FIGURE 1.8. $p_4(x)/q_4(x)$ with grid at $F_5$

```
  ];
  Return[b];
  ]
]

B[k_] := Module[{i, b = IdentityMatrix[2]},
  Which[
  k == 1, Return[{{1, 1}, {1, 0}}],
  k == 2, Return[{{1, 1}, {0, 1}}],
  True,
  For[i = 3, i <= k, i++,
   b = b.{{1, 1}, {0, 1}}
   ];
  Return[b];
  ]
]

Delta[x_, n_] := J[T1[x, n - 1]]
```

```
NearestMediantConvergent[x_, n_] :=
 Module[{i, b = IdentityMatrix[2]},
  For[i = 1, i <= n, i++,
   b = b.B[Delta[x, i]]
   ];
  b
 ]
```

Transformation-based continued fraction representations have been further generalized by Bosma [21] and more recently by Nakada and Natsui [184]

In much of the literature on the Farey series, C. Haros appears as something of a mystery figure. Indeed a tract published by an unnamed institute in 2009 says that Haros' work has "lapsed into oblivion." In the next chapter we hope to convince the reader that C. Haros is alive and well and living in the historical record.

In the literature on continued fractions, there is a second mystery figure living on the edge of the history of mathematics, a certain Dr. Davenant. It was Davenant that sent Wallis a problem concerning the development of a fractional approximation, whose denominator was not greater than 999 that was the starting point of Wallis exploration of the Parmenides algorithm and the theory of continued fractions. Figure 1.9 taken from [234] is Wallis' citation to Dr. Davenant.

40       A L G E B R A.       Cap. X.

## C A P.  X.

### De Fractionum & Rationum Reductione ad minores terminos servato quam potest proxime valore.

PRiusquam hanc de *Decimalibus* doctrinam dimitto, variisque commodis in quotidiana praxi hinc oriundis : libet hic subjungere methodum reducendi Fractiones & Rationes ad minores terminos servato quam potest proxime valore. Quod mihi solvendum Problema, circiter annum si memini 1663 aut 1664, ( per generum suum virum Reverendum D. *Thomam Lamplugh* S. Theologiæ, tum Doctorem, post Episcopum *Exoniensem,* tandem *Eboracensem* Archiepiscopum) misit Reverendus Vir D. *Edwardus Davenant*, S. Theologiæ Doctor, & Ecclesiæ *Sarisburiensis* tum Canonicus Residentiarius ; magnæ eruditionis & modestiæ Vir, & in rebus Mathematicis sedulus, earumque bene gnarus ; mihi saltem fama & per literas notus, scriptisque ex suis aliquot quæ ipso mortuo videre contigit.

Misit ille mihi Fractionem cujus tum Numerator tum Denominator fuerat sex aut septem locorum ; petens, ut huic proxime æqualem exhiberem quæ Denominatorem haberet non majorem quam 999.

Utilitatem hujusmodi inquisitionis hoc specimine conjiciamus.

Rationem Diametri ad Perimetrum Circuli exhibuit *Archimedes*, numeris parvis, ut 7 ad 22 proxime. Nec potest ea numeris non majoribus propius exhiberi.

Quoniam vero res sæpe postulet, ut cum majore accuratione habeatur ea ratio, *Archimedeam* methodum prosecuti sunt alii in majoribus numeris, ad majorem adhuc accurationem. Ut liquet ex *Eutocii* Commentariis in *Archimedis* librum de *Dimensione Circuli.*

Eandem rem prosecuti sunt *Ludovicus Van Culen, Willebrordus Snellius, Adrianus Romanus,* aliique, in numeris saltem locorum 36 ; nescio an majoribus.

Inter alios, *Adrianus Metius,* rem eam prosecutus, rationem illam exhibet ut 113 ad 355 : quæ est, quam *Archimedea* accuratior, minoribus tamen numeris quam *Culenii,* & satis tractabilis; nec potest numeris non majoribus accuratius exhiberi.

Nonnullos, Geometriæ non ignaros, miratos video, qua arte in hanc rationem inciderit *Metius.* Et conjicio quidem id ipsum *Davenantio* contigisse, eumque hac de causa misisse mihi solvendum id Problema. Cujus solutionem ( ante plures annos ) ad ipsum misi ; eamque subjunxi Schediasmatis quibusdam posthumis. *Jeremiæ Horroccii,* quæ meæ curæ commendarunt Societas Regia Londinensis, in ordinem digerenda & edenda. Sed & *Davenantius* ( quod post intellexi ) methodum habuit ipse hujusmodi approximationes, non omnes quidem, sed præcipuas investigandi.

Tractatus mei tum editi ( quia non video quempiam eandem rem ex professo tractasse ) synopsin hic subjiciendam duxi.

### P R O B L E M A.

*Data Fractione* ( seu Ratione ) *quavis, ei quam potest proxime æqualem exhibere, in numeris dato non majoribus, & in minimis terminis.*

Puta, Exposita Fractione $\frac{2684769}{8376571}$ ( seu Ratione 2684769 ad 8376571, ) huic ( si fieri possit) æqualem exhibere, aut ea saltem proxime vel majorem, vel minorem, quæ numeris non-majoribus quam 999 exhiberi potest ; idque in terminis minimis.

### L E M M A.

In ordine ad hanc inquisitionem præmitto, ( tanquam satis notum, aut facile si opus sit demonstratu ) hoc Lemma

*Si Fractionis duo termini ( Numerator & Denominator ) æqualiter utcunque multiplicentur, idem qui prius manet fractionis valor : si vero inæqualiter, valor variatur.*

FIGURE 1.9. The Mysterious Dr. Davenant and His Problem

## 1.12. The Simpson Paradox

Fowler claimed that the Greeks, who reasoned in proportions, might have had difficulty reasoning in decimal notation; we, in turn, who are so thoroughly steeped in decimals, seem to have difficulty reasoning in proportions. This may be so ingrained that we may actually find some of that reasoning paradoxical.

Consider the following line of reasoning:

The death rate of males in the Navy $\left(\frac{3}{5}\right)$ is less than the death rate for males in the Army $\left(\frac{8}{13}\right)$.

The death rate of females in the Navy $\left(\frac{7}{10}\right)$ is less than the death rate for females in the Army $\left(\frac{5}{7}\right)$.

The death rate of soldiers $\left(\frac{8+5}{13+7}\right)$ is less than the death rate for sailors $\left(\frac{3+7}{5+10}\right)$.

To some people that fact that the Navy is safer than the Army sex-by-sex should allow them to conclude that the Navy is safer than the Army full stop. That this conclusion about population statistics doesn't necessarily follow from statistics computed from exhaustive subpopulations was called to the attention of the scientific community by a statistician, Karl Pearson, in a paper in 1899 [192].

Pearson writes about reasoning from correlations and contingency tables, and near the end, he states, "...the reduction of correlation, due to the introduction of fictitious values, is obtained by using as a factor the ratio of actual correlated pairs of individuals to the total number of pairs tabulated." What came to be known as Simpson's paradox was fully elucidated in a paper by George Yule in 1903 [247]. Yule worked with Pearson from 1893 until Pearson's death in 1936.

The fictitious association caused by mixing records finds its counterpart in the spurious correlation to which the same process may give rise in the case of continuous variables, a case to which attention was drawn and which was fully discussed by Professor Pearson in a recent memoir.

Concern about reasoning from subpopulation proportions died down until 1951 when E.H. Simpson published "The Interpretation of Interaction in Contingency Tables" [218]. Simpson raised the same cautions for practitioners of statistics that Pearson and Yule had raised and yet Simpson made no reference to their work.

Things went quiet for another twenty years or so when another applied statistician, Colin Blyth, published a three-page note in the Journal of the American Statistical Association entitled "On Simpson's Paradox and the Sure-Thing Principle" [18]. Beside raising all the statistical cautions again, the paper executed an instance of the Arnol'd' [168] or Boyer-Stigler ([225],[22])Law of Eponymy that no scientific discovery is ever named after its original discoverer. Henceforth, the paradox discovered by Pearson and Yule would be known as Simpson's Paradox.

Blyth's paper set off an avalanche of additional papers by statisticians, logicians, philosophers, and lawyers analyzing the paradox and the nature of what seemed to be a fundamental flaw in human thinking that it highlighted. The fact that the paradox showed up in a sex-bias court case [16] probably contributed to the renewed interest. The final word in this wave of interest in Simpson's Paradox was provided by J.L. Petit in 1992 in his paper "Géneralisation du paradoxe de Simpson" [196].

The core of Simpson's Paradox or as it is also known Simpson's Reversal of Inequalities is a mathematical syllogism whose conclusion is a mediant:

$$a/b < A/B$$
$$c/d < C/D$$
$$(a + c)/(b + d) > (A + C)/(B + D)$$

In 2006 Šleževičienė-Steuding and Steuding published "Simpson's Paradox in the Farey Sequence" [**219**] in which they asked and answered the doubly epony-mous question "How rare is the Simpson reversal in Farey sequences?"

For a fixed pair of fractions $\frac{A}{B}$ and $\frac{C}{D}$ with $\frac{A}{B} < \frac{C}{D}$ in $F_m$ the authors build an elaborate equation for the number of pairs $\frac{a}{c}$ and $\frac{c}{d}$ which satisfy the above syllogism as $m \to \infty$. They refer to this number as $\delta\left(\frac{A}{B}, \frac{C}{D}\right)$. In the paper they give the following examples

$$\delta\left(\frac{1}{2}, \frac{1}{1}\right) = \frac{11}{216} = 0.05092\ldots$$

$$\delta\left(\frac{1}{3}, \frac{2}{3}\right) = \frac{1}{72} = 0.01388\ldots$$

$$\delta\left(\frac{1}{7}, \frac{1}{2}\right) = \frac{1597}{6615} = 0.24142\ldots$$

and conclude with the conjecture based on some numerical experiments that for randomly chosen $\frac{A}{B}$ and $\frac{C}{D}$ in $F_m$ the expected proportion of Simpson reversals goes to 4% as the order of the Farey sequence $m$ goes to infinity, roughly the probability of something of occurring one day in a month.

Figure 1.10 is a plot of $\delta_m\left(\frac{1}{3}, \frac{2}{3}\right)$ for 10(1)200.

```
SimpsonReversals[m_, AB_, CD_] :=
  Module[{f, i, mediant, total = 0, reversals = 0},
   mediant = (Numerator[AB] +
              Numerator[CD])/(Denominator[AB] +
              Denominator[CD]);
   f = Farey[m];
   For[i = 1, i <= Length[f], i++,
    If[f[[i]] >= AB,
    Break[]
    ];
   For[j = 1, j <= Length[f], j++,
    If[f[[j]] >= CD,
     Continue[]
     ];
    total++;
```

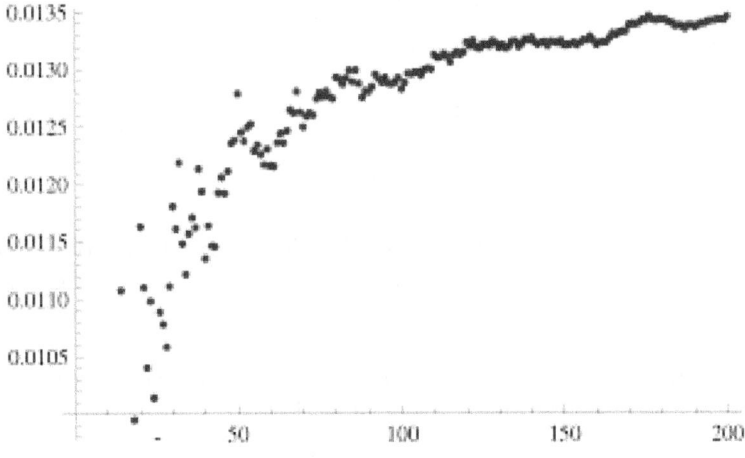

FIGURE 1.10.    $\delta_m \left( \frac{1}{3}, \frac{2}{3} \right)$ for $m = 10(1)200$

```
If[(Numerator[f[[i]]] +
      Numerator[f[[j]]])/(Denominator[f[[i]]] +
      Denominator[f[[j]]]) > mediant,
   reversals++;
   ];
  ];
 ];
reversals/total
];
```

## 1.13. A Motif of Mathematics

The simplicity of the mediant and its ease of use in generating useful sequences and approximations is one of the reasons that it arises in so many many mathematical contexts. A paper by Harold Sinclair Grant, "Additive Entities, an Extension of Farey Series," [113] sets forth all that is needed to have a mediant operation and Farey sequences in a domain. In particular, Grant shows that you only need an operation $+$ that acts on a left-hand entity $A$ and a right-hand entity $B$ with the properties that

(1) $A + (A + B) = (A + A) + B$
(2) $B + A = A + B$
(3) $nA + A = A + nA = (n + 1)A$

Many mathematical constructs and contexts have a combining operation that satisfies these modest conditions and this means that each of them harbors a Farey sequence and, perhaps, a non-arithmetic way of thinking.

In graph theory the mediant is used to study graph colorings. For graph theorists a Farey sequence is the sequence of fractions

$$\frac{k}{1} < \frac{p_n}{q_n} < \ldots < \frac{p_0}{q_0} < \frac{p}{q}$$

generated by the irreducible fraction $\frac{p}{q}$ using the usual Farey condition $p_i q_{i+1} - q_i p_{i+1} = 1$. For example,

$$\left\{ 3, \frac{7}{2}, \frac{11}{3}, \frac{15}{4}, \frac{19}{5}, \frac{23}{6} \right\}$$

```
FareySequenceForGraphTheorists[p_, q_] :=
 Module[{p0 = p, q0 = q, next, l = {p/q}},
  While[q0 > 1,
   next =
    FindInstance[
     p0 q1 - q0 p1 == 1 && p1 < p0 && q1 < q0 && p1 > 0 &&
      q1 > 0, {p1, q1}, Integers];
   p0 = p1 /. next[[1]];
   q0 = q1 /. next[[1]];
   l = Append[l, p0/q0];
   ];
  l
  ]
```

One of the results to which the properties of this Farey sequence contribute is a theorem of Moser [**182**]:

THEOREM 1. *For any rational number r between 2 and 3, there exists a planar graph G whose circular chromatic number is equal to r.*

In functional analysis, Webb [237] harnesses Farey sequences of rational functions. Webb's definition needs to use a little more notation but all of the Farey conditions are easily recognized:

$$\mathcal{F}_n = \{P/Q \mid P, Q \in GF[p, x], \deg(P) < \deg(Q), (P, Q) = 1, Q \text{ monic}\}$$

Webb shows that many of the familiar Farey sequence properties hold and he even gives an algorithm to derive $\mathcal{F}_{n+1}$ from $\mathcal{F}_n$. Webb's mediant is quite a bit more complicated than the sum of numerators divided by sum of denominators since rational equations aren't linearly ordered but it can be used to approximate rational functions in exactly the same way that Chuquet used the règle des nombres moyens to approximate rational numbers.

The mathematical literature also contains numerous generalizations of the Farey sequence itself such as the two-dimensional Farey sequence of Heinrich Made [170] and the roots of integral polynomials of Brown and Mahler [28] but none of these have captured the imagination of practitioners as has the sequence of vulgar fractions.

A long and winding history, utility in diverse applications, accessible for study and elaboration by mathematicians of every stripe, and encoding an enduring mathematical mystery – it is easy to see why David Fowler thought of the mediant as a motif of mathematics.

CHAPTER 2

# History of the Farey Sequence

## 2.1. Mr. R. Flitcon and Question 281

Dickson [**62**] contends that the history of the Farey fractions began at the moment Question 281 appeared in the *The Ladies' Diary for the year of our Lord* 1747. The fateful question, as posed by J. May of Amsterdam [**175**], was the following:

> It is required to find (by a general theorem) the number of fractions of different values, each less than unity, so that the greatest denominator be less than 100?

The first respondent was Mr. Heath, who gave a little hand-crafted table that counted the number of fractions with denominators less than 10 and then finished his submission by saying

> But as there is more trouble than art to discover them, I shall leave it to persons of leisure to pursue the computation, the method being here planned out.

Mr. Ash evidentally wrote down the entire table and proffered the answer 3,055 "but doubts the truth" of it. Mr. Bamfield also completed the table and submitted the answer of 4,851.

In 1751, Mr. R. Flitcon, [**79**] submitted a more general method, at least for denominators less than 100, and using the method arrived at the answer 3,003. Figure 2.1 is Flitcon's solution.

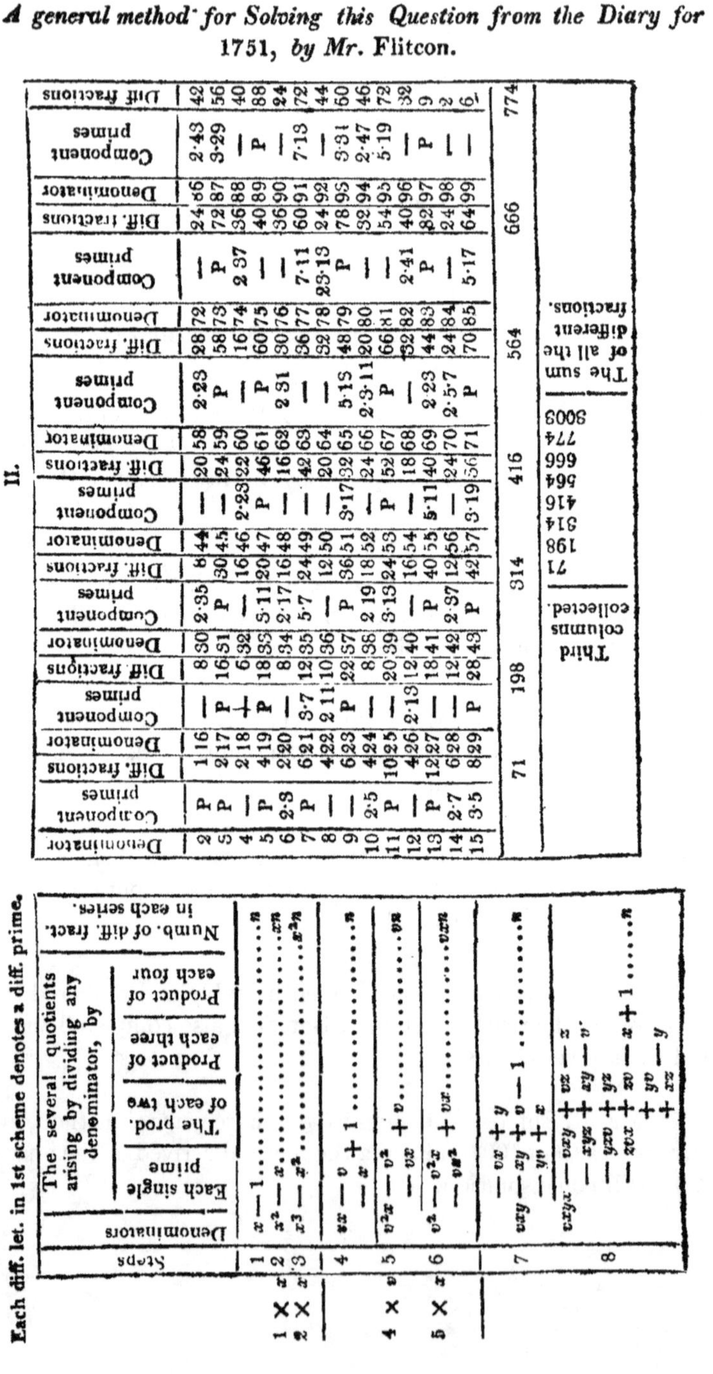

***A general method for Solving this Question from the Diary for 1751, by Mr.* Flitcon.**

II.

| Denominator | Component primes | Diff. fractions | Denominator | Component primes | Diff. fractions | Denominator | Component primes | Diff. fractions | Denominator | Component primes | Diff. fractions | Denominator | Component primes | Diff. fractions | Denominator | Component primes | Diff. fractions | Denominator | Component primes | Diff fractions |
|---|---|---|---|---|---|---|---|---|---|---|---|---|---|---|---|---|---|---|---|---|
| 2 | P | 1 | 16 | — | 8 | 30 | 2·35 | 8 | 44 | — | 20 | 58 | 2·23 | 28 | 72 | — | 24 | 86 | 2·43 | 42 |
| 3 | P | 2 | 17 | P | 16 | 31 | P | 30 | 45 | — | 24 | 59 | P | 58 | 73 | P | 72 | 87 | 3·29 | 56 |
| 4 | — | 2 | 18 | — | 6 | 32 | — | 16 | 46 | 2·23 | 22 | 60 | — | 16 | 74 | 2 37 | 36 | 88 | — | 40 |
| 5 | P | 4 | 19 | P | 18 | 33 | 3·11 | 20 | 47 | P | 46 | 61 | P | 60 | 75 | — | 40 | 89 | P | 88 |
| 6 | 2·3 | 2 | 20 | — | 8 | 34 | 2·17 | 16 | 48 | — | 16 | 62 | 2 31 | 30 | 76 | — | 36 | 90 | — | 24 |
| 7 | P | 6 | 21 | 3·7 | 12 | 35 | 5·7 | 24 | 49 | — | 42 | 63 | — | 36 | 77 | 7·11 | 60 | 91 | 7·13 | 72 |
| 8 | — | 4 | 22 | 2 11 | 10 | 36 | — | 12 | 50 | — | 20 | 64 | — | 32 | 78 | 23·13 | 24 | 92 | — | 44 |
| 9 | — | 6 | 23 | P | 22 | 37 | P | 36 | 51 | 3·17 | 32 | 65 | 5·13 | 48 | 79 | P | 78 | 93 | 3·31 | 60 |
| 10 | 2·5 | 4 | 24 | — | 8 | 38 | 2 19 | 18 | 52 | — | 24 | 66 | 2·3·11 | 20 | 80 | — | 32 | 94 | 2·47 | 46 |
| 11 | P | 10 | 25 | — | 20 | 39 | 3·13 | 24 | 53 | P | 52 | 67 | P | 66 | 81 | — | 54 | 95 | 5·19 | 72 |
| 12 | — | 4 | 26 | 2·13 | 12 | 40 | — | 16 | 54 | — | 18 | 68 | — | 32 | 82 | 2·41 | 40 | 96 | — | 32 |
| 13 | P | 12 | 27 | — | 18 | 41 | P | 40 | 55 | 5·11 | 40 | 69 | 2·23 | 44 | 83 | P | 82 | 97 | P | 9 |
| 14 | 2·7 | 6 | 28 | — | 12 | 42 | 2·37 | 12 | 56 | — | 24 | 70 | 2·5·7 | 24 | 84 | — | 24 | 98 | — | 2 |
| 15 | 3·5 | 8 | 29 | P | 28 | 43 | P | 42 | 57 | 3·19 | 36 | 71 | P | 70 | 85 | 5·17 | 64 | 99 | — | 6 |
| | | 71 | | | 198 | | | 314 | | | 416 | | | 564 | | | 666 | | | 774 |

Third columns collected.

71
198
314
416
564
666
774
3003

The sum of all the different fractions.

**Each diff. let. in 1st scheme denotes a diff. prime.**

| | Steps | Denominators: The several quotients arising by dividing any denominator, by each single prime; the prod. of each two; product of each three; product of each four | Numb. of diff. fract. in each series. |
|---|---|---|---|
| | 1 | $x - 1$ | $n$ |
| $1 \times x$ | 2 | $x^2 - x$ | $xn$ |
| $2 \times x$ | 3 | $x^3 - x^2$ | $x^2n$ |
| | 4 | $vx - v - x + 1$ | $n$ |
| $4 \times v$ | 5 | $v^2x - v^2 - vx + v$ | $vn$ |
| $5 \times x$ | 6 | $v^2 - v^2x - vx^2 + vx$ | $vxn$ |
| | 7 | $vxy - vx - xy - yv + y + v + x - 1$ | $n$ |
| | 8 | $vxyz - vxy - xyz - yzv - zvx + vz + xy + yz + zv + yv + xz - z - v - x - y + 1$ | $n$ |

FIGURE 2.1. Solution to Problem 281 by R. Flitcon, 1751

Flitcon gives the following instructions for the operation of his tables:

N.B. P signifies each prime number in the second column in scheme 2, with a new number to each in the 3d column, deduced from the series in the first scheme, according to the general method: the different denominators being placed in the first column all along.

*For the multipliers, or powers of each.* Against 3 in the 1st column, stands 2 in the 3d column, each being drawn into 3 gives 6 to be set in the 3d column, against 9 in the first collection. Against 6 in the 1st col. is set 2 · 3 in the 2d col. whence (by the 4th step) is found 2 for the 3rd collection. Also for its multiples, into either, or both its parts, as 2 × 3 = 6, drawn into 6 and 2 gives 12 (by the 6th step) to be set in the 3d col. against 36 in the first: dotting through all the second columns all such multiples, by which the fractions are found. Proceeding thus, the whole number of different fractions are truly determined in a short time.

It would seem that Mr. Flitcon was one of those persons of leisure to whom Mr. Heath so graciously bequeathed the problem. Flicton does not employ the mediant but gives a table-based algorithm for sieving out non-vulgar fractions for each denominator. The input to the algorithm is the prime factorization of the denominator.

The equations in the first scheme are the product formulæ for computing Euler's totient function for the specific case that $n$ is less than 100,

$$\phi(n) = n \prod_{p|n} (1 - \frac{1}{p}) = \prod_{p|n} (p-1)p^{k_p-1}.$$

One cannot avoid the suspicion that Mr. Flitcon is playing the magician with the reader and his fellow problem solvers. There's a lot of hocus-pocus in his instructions and it is very likely that he is obfuscating his knowledge of the Euler product in the equations presented in the first scheme. One doesn't come up with the three-prime equation by fitting eight data points. But then hiding your mathematical machinery for use another day is standard practice so hats off to Mr. Flitcon for an elegant solution to Problem 281.

Mr. R. Flitcon's solution did gain him a listing in the *Index of British Mathematicians, Part III, 1701-1800* [**235**], sadly without any additional biographic details.

In case you're wondering if this matter was settled for once and for all by Mr. Flitcon in 1751, the following is a Query in Vol. 4, No. 64 (May, 1907) of *The Mathematical Gazette* by anonymous ENQUIRER:

> Will some correspondent of the *Gazette* publish a list of the principal vulgar fractions $\frac{1}{2}, \frac{1}{3}, \frac{1}{4}, \ldots$ expressed in the forms corresponding to decimals in different scales of notation? The list need not be a long one; it might go up to $\frac{1}{12}$ and 12 as the base, or better up to $\frac{1}{20}$ with 20 as the base, an it might be conveniently put in a tabular form. I do not know if such as list has been published anywhere. Might it not be worth reprinting in the *Gazette*? It has been said that the decimal system of notation is better than any other, and such a list would surely afford some test of this point.

Responses to such queries presumably started coming faster after May 8, 1962 when U.S. Patent 3,033,451 was issued. The inventor is Gaetano Badalameti of Via Adda 13, Suisio, Bergamo, Italy. The title of the patent is "Slide Rule for Ascertaining the Divisibility of Numbers and for the Resolution into Prime Number Factors of Such Numbers." The diagram of Mr. Badalamenti's slide rule taken from his patent is shown as Figure 2.2.

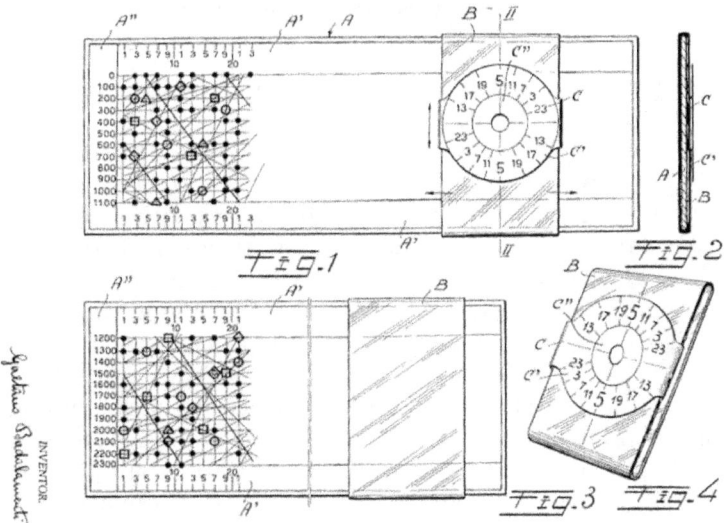

FIGURE 2.2. US Patent 3,033,451 - Factoring Slide Rule

## 2.2. Charles Haros, Géomètre

One of the most prescient and enduring of the changes introduced by the French Revolution was the metric system of mensuration. The metric system was legislated in 1791 but there was little momentum for enforcing its use until the official meter and kilogram were deposited in Archives of the Republic on June 22, 1799. From this date on the government actively encouraged the public to abandon the fraction and learn to love the decimal. Unlike switching from driving on the right side of the street to driving on the left side, however, switching from fractional notation to decimal notation would entail a considerable education effort, and couldn't take place all at once.

As part of the education effort a géomètre at the Bureau du Cadastre de Paris, Charles Haros, created a set of tables for converting between fractional and decimal notation. Regarding Haros' job title, E.H. Neville, who will play a rôle in this history later on, notes the following in the Gleanings Far and Near section of Vol. 14, No. 203 (October, 1929) of *The Mathematical Gazette*:

> It is to be remembered that 'géomètre' in French is no narrower than 'mathematician' in English ...

In fact Napoleon referred to Laplace as a géomètre in the process of firing him as his Minister of the Interior six weeks after appointing him to the job:

> Géomètre de premier rang, Laplace ne tarda pas á se montrer administrateur plus que médiocre; dès son premier travail nous reconnûmes que nous étions trompé.

To build his tables Haros needed to create the list of all irreducible fractions between 0 and 1 with denominators between 2 and 99. There is no evidence that he knew how many such fractions there were before building the list. It is unlikely that he read back issues of *The Ladies Diary*. But he surely knew that just writing down irreducible fractions with these denominators as he thought of them wasn't the right way of going about the task. He needed an algorithm, along with a convincing argument that it delivered all the required fractions.

Haros contributed a paper to the Annonces et Notices section of Volume IV, Eleventh Notebook, of the *Journal de l'École Polytechnique* [**128**] published in the month of Messidor of Year X. The paper not only gave the algorithm but also a sketch of a proof that it worked. While the paper may have purported to be written to inform the public at large, it is a little hard to believe that a revolutionary street vendor required the mathematical proofs backing up the tables before using them. One might also wonder how many shop keepers subscribed to the *Journal de l'École Polytechnique*.

Setting aside the purpose of the paper, the purpose of the tables was to provide everyone with a handy method for converting numbers written in the fractional representation of the *ancien régime* to their revolution-decreed decimal representation and for converting numbers written in the decimal representation to an approximate fractional representation. It was in constructing a table to facilitate this latter conversion from a decimal representation to an approximating simple fraction, that what become known as the Farey sequence made its first appearance.

Immediately below is the original paper by Charles Haros followed by a translation to which has been interpolated examples of what the text is describing.

## 2.3. "Tables pour évaluer une fraction ordinaire ... "

TABLES *pour évaluer une fraction ordinaire avec autant de décimales qu'on voudra; et pour trouver la fraction ordinaire la plus simple, et qui approche sensiblement d'une fraction décimale.* Par le C.*en* HAROS.

Pour évaluer une fraction ordinaire en decimales, on sait qu'il faut diviser le numérateur par le dénominateur, en observant d'ajouter successivement un zéro à chaque reste de division, et continuer ainsi jusqu'à ce qu'on ait un quotient exact ou une valuer suffisamment approchée.

Lorsque le dénominateur d'une fraction irréducible est un nombre impair non divisible par 5, les restes de division sont irréducibles par rapport au diviseur; car le zéro que l'on ajoute successivement à chaque reste, donne un dividénde 10 fois plus grand : or, le nombre 10 n'a d'aprés l'hypothése, aucun facteur commun avec le diviseur ; donc, aprés la division, le reste du dividende partiel ne peut avoir aucun facteur commun avec le diviseur. La division étant poussée suffisamment loin, donne des restes égaux à ceux qu'on a déjà trouvés ; d'ou il suit que les chiffres du quotient reparaissent de nouveau, et forment une période. Cette période ne peut être composée de plus de chiffres au quotient, qu'il y a d'unités moins une dans le diviseur ou dénominateur. Il est des cas où le diviseur, quoique très-grand, donne néanmoins une période très-petite.

On çoncoit en effet qu'en poussant la division, il doit arriver de deux choses l'une; où tous les restes seront différens, et dans ce cas leur nombre ne peut surpasser le diviseur moins un, parce que ces restes sont chacun plus petits que le diviseur; ou dans le cours de la division on trouvera un reste égal à un des précédens, alors les mêmes chiffres du quotient reparaîtront. On pourra donc, dans ce cas, écrire, à la suite des chiffres déjà trouvés au quotient, tel nombre de décimales qu'on voudra, sans avoir la peine de continuer la division.Enfin, si l'on considère les restes successifs des divisions partielles comme numérateurs,ils formeront avec le diviseur autant de fractions irréducibles qu'il y aura de chiffres dans la période; et la valeur de chacune de ces fractions, exprimée en dixièmes seulement, sera égale au chiffre corresondant du quotient.

Quand le dénominateur d'une fraction irréductible est un nombre pair, et que la division ne peut se faire exactement, les chiffres du quotient ne sont qu'en partie périodiques, c'est-à-dire, que l'on trouve dans le commencement de la division un ou plusieurs chiffres non périodiques; les restes de division sont tous des nombres pairs, et par conséquent susceptibles d'être réduits avec le diviseur à des termes plus simples.

C'est d'après ces propriétés que j'ai calculé une table d'une nouvelle disposîtion pour évaluer une fraction irréductible dont le dénominateur ne surpasse pas 99, avec autant de décimales qu'on voudra. Voici la manière de s'en servir.

Lorsqu'on aura une fraction ordinaire à évaluer en décimales, on cherchera le dénominateur en tête des colonnes, et le numérateur dans la petite colonne de gauche ; on trouvera dans celle de droite, et vis-à-vis du numérateur, la valeur exacte ou approchée de la fraction ordinaire. Si cette valeur est suivie du mot abrégé *ex.*, on aura la valeur exacte de la fraction en décimales. Si la partie décimale se trouve terminée par &c., la dèrniere décimale doit être alors répetée autant de fois qu'on voudra. Si l'on ne trouve aucune expression après la quantité décimale, on prendra cette valeur telle qu'elle se trouve dans la colonne, et de suite on écrira à sa droite tous les chiffres, après les virgules, que l'on trouvera en descendant le long de la colonne: arrivé à la lettre P, qui signifie période, on pourra, au besoin, prendre d'autres chiffres en tête de la période, en observant de ne pas dépasser le trait qui en marque la limite. Enfine, si la partie décimale est accompagnée d'une fraction ordinaire, on écrira à la suite de cette partie la valeur de cette fraction.

Soit propose d'evaluer $\frac{7}{16}$ en décimales. Je cherche 16 en tête des colonnes, et 7 dans la petite colonne de gauche: je trouve dans celle de droite, 0,4375 pour la valeur exacte de $\frac{7}{16}$.

Soit $\frac{17}{36}$ à évaluer en décimales. Après avoir trouvé la colonne en tête de laquelle est 36, et le nombre 17 dans la petite colonne, je trouve , à côté de ce dernier nombre, 0,472 &c., c'est-à-dire, 0,4722222 &c., pour la valeur de $\frac{17}{36}$.

Proposons-nous d'avoir la valeur de $\frac{19}{47}$ avec vingt décimales. Je cherche 47 en t̂ete des colonnes, et je trouve que 19 répond à 0,4. Maintenant j'écris de suite, à c̊oté de cette valeur, tous les chiffres 0.4.2.5.5.3, &c., que je trouve en descendant; j'ai 0,40425531914893617. Comme je n'ai encore que 17 décimales, je prends les trois autres en tête de la période, c'est-à-dire, 0.2.1, et j'ai 0,40425531914893617021 pour la valer de $\frac{19}{47}$.

Soit proposé, pour dernier exemple, d'évaluer $\frac{31}{95}$ en décimales. Je trouve pour valuer $0,3\frac{5}{19}$. Je cherche dans un autre colonne la valeur de $\frac{5}{19}$; je trouve 0,2. J'écris d'abord le 2 à côté de 0,3 , et de suite les chiffres 6.3.1.5.7.8.9, &c. J'ai 0,326315789473 pour la valeur de $\frac{31}{95}$.

Pour trouver une fraction ordinaire qui approche sensiblement d'une quantité décimale, on considère cette dernière quantité comme un nombre entier à diviser par l'unité, suivie d'autant de zéros qu'il y a de décimales. La division n'étant qu'indiquée, on a une fraction que l'on réduit en fraction continue ; alors les fractions intégrantes donnent des fractions ordinaires qui approchent sensiblement de la fraction décimale.

Comme ces calculs sont un peu longs, je construis dans ce moment une table, au moyen de laquelle, étant donnée une fraction décimale, on pourra trouver sur le champ la fraction ordinaire la plus simple qui en approchera le plus. Pour cet effet, il suffira de disposer les 3003 fractions qui résultent de la table précédente, par ordre de grandeurs; et pour y parvenir directement, je me propose le problème suivant:

Trouver, par ordre de grandeurs, toutes les fractions irréductibles comprises entre 0 et 1, avec la condition que les dénominateurs de ces fractions ne surpassent pas deux chiffres.

Ce problème résolu, il n'y aura plus qu'à écrire , à côté de chaque fraction, sa valeur en décimales que l'on trouvera dans la première table.

Pour résoudre ce problème, j'écris d'abord cette suite de fractions:

$$\frac{1}{99} \cdot \frac{1}{98} \cdot \frac{1}{97} \cdots \frac{1}{4} \cdot \frac{1}{3} \cdot \frac{1}{2} \cdot \frac{2}{3} \cdot \frac{3}{4} \cdots \frac{96}{97} \cdot \frac{97}{98} \cdot \frac{98}{99};$$

dans laquelle une fraction quelconque diffère de sa voisine, de l'unité divisée par le produit de leurs dénominateurs. Cette propriété s'apercoit mieux dans la première moitié de la suite que dans la seconde : mais si on représente une fraction quelconque de la seconde moitié par $\frac{(a-1)}{a}$ , la fraction suivante sera $\frac{a}{(a+1)}$; et on aura, pour la différence de ces deux fractions, $\frac{1}{a(a+1)}$, c'est-à-dire, l'unité divisée par le produit de leurs dénominateurs.

Il nous reste maintenant à intercaler entre les fractions précédentes, toutes celles qui ont des dénominateurs moindres que 100, et qui sont irréductibles.Il est necéssaire, dans cette recherche, que les fractions intermediaires se suivent par ordre de grandeurs , et que la différence d'une fraction à sa voisine soit toujours égale a l'unité divisee par le produit de leurs dénominateurs ; parce qu'alors une fraction quelconque de la suite sera irréductible, et exprimera de la manière la plus simple la valeur approchée de l'une ou de l'autre des deux fractions entre lesquelles elle se trouvera.

Soient $\frac{a}{b}$ et $\frac{c}{d}$ deux fractions telles que $bc - ad = 1$; et proposons-nous de déterminer une fraction intermédiaire $\frac{x}{y}$, telle qu'on ait $bx - ay = 1$ et $cy - dx = 1$. En résolvant ces deux équations, on trouvera $x = \frac{(a+c)}{(bc-ad)}$ et

$y = \frac{(b+d)}{(bc-ad)}$; mais, d'après l'hypothèse, $bc - ad = 1$ ; donc $x = a + c$, et $y = b + d$; par conséquent $\frac{x}{y} = \frac{(a+c)}{(b+d)}$. Ce résultat nous apprend que la fraction intermédiaire est égale à la somme des numérateurs des fractions $\frac{a}{b}$ et $\frac{c}{d}$, divisee par la somme des dénominateurs. Comme les trois fractions $\frac{a}{b}$, $\frac{(a+c)}{(b+d)}$, $\frac{c}{d}$ diffèrent entre elles de $bc-ad = 1$, divisée par le produit de leur dénominateur, la fraction intermédiaire est donc irréductible, et se trouve en même temps la fraction la plus simple qui approche le plus de l'une ou de l'autre des deux fractions $\frac{a}{b}$ et $\frac{c}{d}$.

Les dénominateurs des fractions qui doivent composer la table en question, étant moindres que 100, on ne pourra intercaler aucune fraction dans la suite $\frac{1}{99} \cdot \frac{1}{98} \cdot \frac{1}{97} \cdots \frac{1}{50}$, parce que la somme des dénominateurs de deux fractions voisines surpasse 99 mais, entre $\frac{1}{50}$ et $\frac{1}{49}$, on pourra intercaler $\frac{2}{99}$. Les fractions $\frac{1}{49}$ et $\frac{1}{48}$ donneront $\frac{2}{97}$; et l'on peut aller ainsi de suite jusqu'à la fraction $\frac{1}{33}$; alors, entre celle-ci et $\frac{1}{32}$, il sera possible d'intercaler trois fractions; on aura d'abord $\frac{1}{33} \cdot \frac{2}{65} \cdot \frac{1}{32}$, puis $\frac{1}{33} \cdot \frac{3}{98} \cdot \frac{2}{65} \cdot \frac{3}{97} \cdot \frac{1}{32}$.

D'après ces exemples, on doit voir comment on peur trouver toutes les fractions intermédiaires entre $\frac{1}{32}$ et $\frac{1}{31}$, entre $\frac{1}{31}$ et $\frac{1}{30}$; et ainsi de suite.

## 2.4. "Tables for evaluating a common fraction ..."

TABLES *for evaluating a common fraction to as many decimal points as one wishes; and for determining the simplest common fraction that closely approximates a decimal fraction.* By C·*en* HAROS.

To evaluate a common fraction in decimals, we all know that we must divide the numerator by the denominator, taking care to add an additional zero to each remainder in the division, and to continue in this fashion until obtaining an exact quotient or a value that is sufficiently close.

When the denominator of an irreducible fraction is an odd number not divisible by 5, the remainders of the division are irreducible with respect to the divisor; since the zero that one adds successively to each remainder yields

TABLE 2.1. Fraction to Decimal Computation

| 3 | | | | | | $/7 = 0.$ |
|---|---|---|---|---|---|---|
| 3 | 0 | | | | | $/7 = 4$ |
| | 2 | 0 | | | | $/7 = 2$ |
| | | 6 | 0 | | | $/7 = 8$ |
| | | | 4 | 0 | | $/7 = 5$ |
| | | | | 5 | 0 | $/7 = 7$ |
| | | | | | 1 | 0 | $/7 = 1$ |
| | | | | | | 5 | etc. |

a dividend that is 10 times as great; now according to the hypothesis, the number 10 has no common factor with the divisor; thus, after the division, the remainder of the partial dividend cannot have any common factor with the divisor. If the division is carried forward sufficiently, the remainders will be the same as those already found; from which it follows that the numbers in the quotient will reappear once again, and form a period. This period may not be composed of more digits in the quotient than one less than the number of digits in the divisor or denominator. There are instances where the divisor, although very large, results in a period that is very small.

One can understand that, in fact, when carrying forward the division, one of two things must occur: one where all the remainders will be different, and in this case their number cannot exceed the divisor minus one, because these remainders are each smaller than the divisor; or over the course of the division, one will find a remainder that is equal to one of the preceding ones, and then the same figures of the quotient will reappear. In this case, one may thus continue writing the same numbers already found in the quotient for as many decimal places as one wishes, without having to take the trouble of continuing the division process. Finally, if one considers the successive remainders of the partial divisions as numerators, then, together with the divisor, they will form as many irreducible fractions as there are digits in the period; and, the value of each of these fractions, expressed only in tenths, will be equal to the corresponding number in the quotient.

When the denominator of an irreducible fraction is an even number, and the division cannot be performed exactly, the numbers in the quotient are only partially periodic; that is to say, in the beginning of the division one finds one or several numbers that are non-periodic; the remainders of the division are all even numbers, and as a consequence, able to be reduced by the divisor to simpler terms.

It is in accordance with these properties that I have calculated a new kind of table for evaluating an irreducible fraction whose denominator does not exceed 99 to as many decimal places that one wishes. Here is the way to use it.

When you have an common fraction to transform into decimals, you find the denominator at the head of the columns, and the numerator in the small column on the left. On the column located to the right, just opposite the numerator, you can find the exact value or an approximation of the common fraction. If this value is followed by the abbreviation "ex," you will obtain the exact value of the fraction in decimals. If the decimal portion terminates in "&c." then the last decimal should then be repeated as many times as you wish. If there is no expression following the decimal quantity, then you should take this value as it is given in the column, and then you should write all the numbers to its right that follow the commas you will find while going down the length of the column. If you come to the letter P, which signifies "period," you may, if necessary, take other figures ahead of the period, while taking care not to exceed the character that marks its limit. Finally, if the decimal part is accompanied by a common fraction, you will write the value of this fraction after this part.

Suppose that the task is to evaluate $\frac{7}{16}$ in decimals. I look for 16 at the head of the columns, and for 7 in the small column on the left. In the column on the right, I find 0.4375 as the exact value for $\frac{7}{16}$.

Suppose that the task is to evaluate $\frac{17}{36}$ in decimals. After finding the column headed by the number 36, and the number 17 in the small column, I find, next to the latter number, 0.472 etc., that is to say, 0.4722222 etc., as the value of $\frac{17}{36}$.

Let us propose calculating the value of $\frac{19}{47}$ to twenty decimal places. I look for 47 at the head of the columns, and I find that 19 corresponds to 0.4. Now, I write next to this value all of the numbers 0.4.2.5.5.3, etc., that I find going

TABLE 2.2. Fraction to Decimal Table - Exact

| 16 | |
|---|---|
| 1 | 0.0625 *ex.* |
| 3 | 0.1875 *ex.* |
| 5 | 0.3125 *ex.* |
| 7 | 0.4375 *ex.* |
| 9 | 0.5625 *ex.* |
| 11 | 0.6875 *ex.* |
| 13 | 0.8125 *ex.* |
| 15 | 0.9375 *ex.* |

TABLE 2.3. Fraction to Decimal Table - Repeated

| 36 | |
|---|---|
| 1 | 0.027 *etc.* |
| 5 | 0.138 *etc.* |
| 7 | 0.194 *etc.* |
| 11 | 0.305 *etc.* |
| 13 | 0.361 *etc.* |
| 17 | 0.472 *etc.* |
| 19 | 0.527 *etc.* |
| 23 | 0.638 *etc.* |
| 25 | 0.694 *etc.* |
| 29 | 0.805 *etc.* |
| 31 | 0.861 *etc.* |
| 35 | 0.972 *etc.* |

down the column: I have 0.40425531914893617. Since I only have 17 decimal places, I find the next three ahead of the period, that is to say, 0.2.1, and I have 0.40425531914893617021 for the value of $\frac{19}{47}$.

TABLE 2.4. Fraction to Decimal Table - Periodic

| 47 | | 47 | |
|---|---|---|---|
| 1 | 0.0 | 46 | 0.9 |
| 10 | 0.2 | 37 | 0.7 |
| 6 | 0.1 | 41 | 0.8 |
| 13 | 0.2 | 34 | 0.7 |
| 36 | 0.7 | 11 | 0.2 |
| 31 | 0.6 | 16 | 0.3 |
| 28 | 0.5 | 19 | 0.4 |
| 45 | 0.9 | 2 | 0.0 |
| 27 | 0.5 | 20 | 0.4 |
| 35 | 0.7 | 12 | 0.2 |
| 21 | 0.4 | 26 | 0.5 |
| 22 | 0.4 | 25 | 0.5 |
| 32 | 0.6 | 15 | 0.3 |
| 38 | 0.8 | 9 | 0.1 |
| 4 | 0.0 | 43 | 0.9 |
| 40 | 0.8 | 7 | 0.1 |
| 24 | 0.5 | 23 | 0.4 |
| 5 | 0.1 | 42 | 0.8 |
| 3 | 0.0 | 44 | 0.9 |
| 30 | 0.6 | 17 | 0.3 |
| 18 | 0.3 | 29 | 0.6 |
| 39 | 0.8 | 8 | 0.1 |
| 14 | 0.2 | 33 | 0.7$P$ |

TABLE 2.5. Fraction to Decimal Table - Decimal with Fraction

| 95 | | | 95 | | | 95 | | | 95 | | |
|---|---|---|---|---|---|---|---|---|---|---|---|
| 1 | 0.0 | $\frac{2}{19}$ | 24 | 0.2 | $\frac{10}{19}$ | 48 | 0.5 | $\frac{1}{19}$ | 72 | 0.7 | $\frac{11}{19}$ |
| 2 | 0.0 | $\frac{4}{19}$ | 26 | 0.2 | $\frac{14}{19}$ | 49 | 0.5 | $\frac{3}{19}$ | 73 | 0.7 | $\frac{13}{19}$ |
| 3 | 0.0 | $\frac{6}{19}$ | 27 | 0.2 | $\frac{16}{19}$ | 51 | 0.5 | $\frac{7}{19}$ | 74 | 0.7 | $\frac{15}{19}$ |
| 4 | 0.0 | $\frac{8}{19}$ | 28 | 0.2 | $\frac{18}{19}$ | 52 | 0.5 | $\frac{9}{19}$ | 77 | 0.8 | $\frac{2}{19}$ |
| 6 | 0.0 | $\frac{12}{19}$ | 29 | 0.3 | $\frac{1}{19}$ | 53 | 0.5 | $\frac{11}{19}$ | 78 | 0.8 | $\frac{4}{19}$ |
| 7 | 0.0 | $\frac{14}{19}$ | 31 | 0.3 | $\frac{5}{19}$ | 54 | 0.5 | $\frac{13}{19}$ | 79 | 0.8 | $\frac{6}{19}$ |
| 8 | 0.0 | $\frac{16}{19}$ | 32 | 0.3 | $\frac{7}{19}$ | 56 | 0.5 | $\frac{17}{19}$ | 81 | 0.8 | $\frac{10}{19}$ |
| 9 | 0.0 | $\frac{18}{19}$ | 33 | 0.3 | $\frac{9}{19}$ | 58 | 0.6 | $\frac{2}{19}$ | 82 | 0.8 | $\frac{12}{19}$ |
| 11 | 0.1 | $\frac{3}{19}$ | 34 | 0.3 | $\frac{11}{19}$ | 59 | 0.6 | $\frac{4}{19}$ | 83 | 0.8 | $\frac{14}{19}$ |
| 12 | 0.1 | $\frac{5}{19}$ | 36 | 0.3 | $\frac{15}{19}$ | 61 | 0.6 | $\frac{8}{19}$ | 84 | 0.8 | $\frac{16}{19}$ |
| 13 | 0.1 | $\frac{7}{19}$ | 37 | 0.3 | $\frac{17}{19}$ | 62 | 0.6 | $\frac{10}{19}$ | 86 | 0.9 | $\frac{1}{19}$ |
| 14 | 0.1 | $\frac{9}{19}$ | 39 | 0.4 | $\frac{2}{19}$ | 63 | 0.6 | $\frac{12}{19}$ | 87 | 0.9 | $\frac{3}{19}$ |
| 16 | 0.1 | $\frac{13}{19}$ | 41 | 0.4 | $\frac{6}{19}$ | 64 | 0.6 | $\frac{14}{19}$ | 88 | 0.9 | $\frac{5}{19}$ |
| 17 | 0.1 | $\frac{15}{19}$ | 42 | 0.4 | $\frac{8}{19}$ | 66 | 0.6 | $\frac{18}{19}$ | 89 | 0.9 | $\frac{7}{19}$ |
| 18 | 0.1 | $\frac{17}{19}$ | 43 | 0.4 | $\frac{10}{19}$ | 67 | 0.7 | $\frac{1}{19}$ | 91 | 0.9 | $\frac{11}{19}$ |
| 21 | 0.2 | $\frac{4}{19}$ | 44 | 0.4 | $\frac{12}{19}$ | 68 | 0.7 | $\frac{3}{19}$ | 92 | 0.9 | $\frac{13}{19}$ |
| 22 | 0.2 | $\frac{6}{19}$ | 46 | 0.4 | $\frac{16}{19}$ | 69 | 0.7 | $\frac{5}{19}$ | 93 | 0.9 | $\frac{15}{19}$ |
| 23 | 0.2 | $\frac{8}{19}$ | 47 | 0.4 | $\frac{18}{19}$ | 71 | 0.7 | $\frac{9}{19}$ | 94 | 0.9 | $\frac{17}{19}$ |

Suppose that the final example requires evaluating $\frac{31}{95}$ in decimals. I find a value of 0.3 $\frac{5}{19}$. I look up the value of $\frac{5}{19}$ in another column, and find that it is 0.2. I first write the 2 next to 0.3, and then the numbers, 6.3.1.5.7.8.9, etc. I come up with 0.326315789473 as the value of $\frac{31}{95}$.

In order to find a common fraction that reasonably approximates a decimal quantity, you will consider the decimal quantity as a whole number divided by one, followed by as many zeros as there are decimal places. Since the division is only implied, one has a fraction that one reduces into a continuous fraction; thus, the integrated fractions result in common fractions that reasonably approximate the decimal fraction.

Since these calculations are a bit long, at this moment I construct a table, by means of which, given a decimal fraction, you can immediately find the simplest common fraction that will best approximate it. For this purpose, it is

TABLE 2.6. Fraction to Decimal Table - Periodic

| 19 | | 19 | |
|---|---|---|---|
| 1 | 0.0 | 18 | 0.9 |
| 10 | 0.5 | 9 | 0.4 |
| 5 | 0.2 | 14 | 0.7 |
| 12 | 0.6 | 7 | 0.3 |
| 6 | 0.3 | 13 | 0.6 |
| 3 | 0.1 | 16 | 0.8 |
| 11 | 0.5 | 8 | 0.4 |
| 15 | 0.7 | 4 | 0.2 |
| 17 | 0.8 | 2 | 0.1$P$ |

sufficient to have 3003 fractions available that emerge from the preceding table, arranged according to magnitude, and in order to arrive at it directly, I assign myself the following problem:

To find, in order of magnitude, all the irreducible fractions ranging between 0 and 1, meeting the condition that the denominators of these fractions not exceed two digits.

Once this problem is resolved, you will merely have to write next to each fraction its value in decimals that you will find in the first table.

To resolve this problem, I first write this series of fractions:

$$\frac{1}{99}, \frac{1}{98}, \frac{1}{97} \cdots \frac{1}{4}, \frac{1}{3}, \frac{1}{2}, \frac{2}{3}, \frac{3}{4} \cdots \frac{96}{97}, \frac{97}{98}, \frac{98}{99},$$

in which every fraction differs from its neighbor by one divided by the product of their denominators. This property can be better appreciated in the first half of the series than in the second, but if you represent any fraction in the second half by $\frac{(a-1)}{a}$, the next fraction will be $\frac{a}{(a+1)}$; and for the difference between these two fractions, you will have $\frac{1}{a(a+1)}$, that is to day, one divided by the product of their denominators.

Now, we only are left to intercalate between the preceding fractions all those that have denominators less than 100 and are irreducible. In this search, it is necessary that the intermediate fractions are arranged according to their magnitude, and that the difference between one fraction and its neighbor will always be equal to one divided by the product of their denominators, since then any fraction along the way will be irreducible and will express in the simplest possible way the value approached by one or the other of the two fractions between which it is located.

Let $\frac{a}{b}$ and $\frac{c}{d}$ be two fractions such that $ad - bc = 1$; and let us propose to determine an intermediate fraction $\frac{x}{y}$ such that one has $bx - ay = 1$ and $cy - dx = 1$. In solving these two equations, one will find that $x = \frac{(a+c)}{(bc-ad)}$ and that $y = \frac{(b+d)}{(bc-ad)}$; however, according to the hypothesis, $ad - bc = 1$, and thus, $x = a+c$, and $y = b+d$; as a consequence, $\frac{x}{y} = \frac{(a+c)}{(b+d)}$. This result shows us that the intermediate fraction is equal to the sum of the numerators of the fractions $\frac{a}{b}$ and $\frac{c}{d}$ divided by the sum of the denominators. Since the three fractions $\frac{a}{b}$, $\frac{(a+c)}{(b+d)}$, and $\frac{c}{d}$ differ between themselves by $bc - ad = 1$ divided by the product of their denominator, the intermediate fraction is thus irreducible, and at the same time represents the simplest fraction that most closely approaches one or the other of the two fractions $\frac{a}{b}$ and $\frac{c}{d}$.

Since the denominators of fractions to be included in the table in question must be less than 100, you cannot intercalate any fraction in the series $\frac{1}{99} \cdot \frac{1}{98} \cdot \frac{1}{97} \cdots \frac{1}{50}$ because the sum of the denominators of two neighboring fractions will exceed 99. However, between $1/5\frac{1}{50}$ and $\frac{1}{49}$, you can intercalate $\frac{2}{99}$. The fractions $\frac{1}{49}$ and $\frac{1}{48}$ will yield $\frac{2}{97}$; and you can proceed in this way up to the fraction $\frac{1}{33}$; then, between the latter and $\frac{1}{32}$, it will be impossible to intercalate three fractions; you will first have $\frac{1}{33} \cdot \frac{2}{65} \cdot \frac{1}{32}$ then $1/33.3/98.2/65.3/97.1/32$.

From these examples, you can readily see how it is possible to find all the intermediate fractions between $\frac{1}{32}$ and $\frac{1}{31}$, between $\frac{1}{31}$ and $\frac{1}{30}$, ; and so forth.

## 2.5. The Farey Sequence as the Argument of a Mathematical Table

In the second, decimal-to-fraction part of the paper, Haros builds a table of all irreducible fractions with denominators less than or equal to 99 and then uses the fraction-to-decimal tables from the first part of the paper to associate a decimal representation with each fraction. In order to convert from a decimal

TABLE 2.7. A Middle Part of Haros' Decimal-to-Fraction Table

| | | | | | | | |
|---|---|---|---|---|---|---|---|
| $\frac{12}{25}$ | 0.480000 | $\frac{45}{92}$ | 0.489130 | $\frac{1}{2}$ | 0.500000 | $\frac{47}{92}$ | 0.510870 |
| $\frac{37}{77}$ | 0.480519 | $\frac{23}{47}$ | 0.489362 | $\frac{50}{99}$ | 0.505051 | $\frac{23}{45}$ | 0.511111 |
| $\frac{25}{52}$ | 0.480769 | $\frac{47}{96}$ | 0.489583 | $\frac{49}{97}$ | 0.505155 | $\frac{45}{88}$ | 0.511364 |
| $\frac{38}{79}$ | 0.481013 | $\frac{24}{49}$ | 0.489796 | $\frac{48}{95}$ | 0.505263 | $\frac{22}{43}$ | 0.511628 |
| $\frac{13}{27}$ | 0.481481 | $\frac{25}{51}$ | 0.490196 | $\frac{47}{93}$ | 0.505376 | $\frac{43}{84}$ | 0.511905 |
| $\frac{40}{83}$ | 0.481928 | $\frac{26}{53}$ | 0.490566 | $\frac{46}{91}$ | 0.505495 | $\frac{21}{41}$ | 0.512195 |
| $\frac{27}{56}$ | 0.482143 | $\frac{27}{55}$ | 0.490909 | $\frac{45}{89}$ | 0.505618 | $\frac{41}{80}$ | 0.512500 |
| $\frac{41}{85}$ | 0.482353 | $\frac{28}{57}$ | 0.491228 | $\frac{44}{87}$ | 0.505747 | $\frac{20}{39}$ | 0.512821 |
| $\frac{14}{29}$ | 0.482759 | $\frac{29}{59}$ | 0.491525 | $\frac{43}{85}$ | 0.505882 | $\frac{39}{76}$ | 0.513158 |
| $\frac{43}{89}$ | 0.483146 | $\frac{30}{61}$ | 0.491803 | $\frac{42}{83}$ | 0.506024 | $\frac{19}{37}$ | 0.513514 |
| $\frac{29}{60}$ | 0.483333 | $\frac{31}{63}$ | 0.492063 | $\frac{41}{81}$ | 0.506173 | $\frac{37}{72}$ | 0.513889 |
| $\frac{44}{91}$ | 0.483516 | $\frac{32}{65}$ | 0.492308 | $\frac{40}{79}$ | 0.506329 | $\frac{18}{35}$ | 0.514286 |
| $\frac{15}{31}$ | 0.483871 | $\frac{33}{67}$ | 0.492537 | $\frac{39}{77}$ | 0.506494 | $\frac{35}{68}$ | 0.514706 |
| $\frac{46}{95}$ | 0.484211 | $\frac{34}{69}$ | 0.492754 | $\frac{38}{75}$ | 0.506667 | $\frac{17}{33}$ | 0.515152 |
| $\frac{31}{64}$ | 0.484375 | $\frac{35}{71}$ | 0.492958 | $\frac{37}{73}$ | 0.506849 | $\frac{50}{97}$ | 0.515464 |
| $\frac{47}{97}$ | 0.484536 | $\frac{36}{73}$ | 0.493151 | $\frac{36}{71}$ | 0.507042 | $\frac{33}{64}$ | 0.515625 |
| $\frac{16}{33}$ | 0.484848 | $\frac{37}{75}$ | 0.493333 | $\frac{35}{69}$ | 0.507246 | $\frac{49}{95}$ | 0.515789 |
| $\frac{33}{68}$ | 0.485294 | $\frac{38}{77}$ | 0.493506 | $\frac{34}{67}$ | 0.507463 | $\frac{16}{31}$ | 0.516129 |
| $\frac{17}{35}$ | 0.485714 | $\frac{39}{79}$ | 0.493671 | $\frac{33}{65}$ | 0.507692 | $\frac{47}{91}$ | 0.516484 |
| $\frac{35}{72}$ | 0.486111 | $\frac{40}{81}$ | 0.493827 | $\frac{32}{63}$ | 0.507937 | $\frac{31}{60}$ | 0.516667 |
| $\frac{18}{37}$ | 0.486486 | $\frac{41}{83}$ | 0.493976 | $\frac{31}{61}$ | 0.508197 | $\frac{46}{89}$ | 0.516854 |
| $\frac{37}{76}$ | 0.486842 | $\frac{42}{85}$ | 0.494118 | $\frac{30}{59}$ | 0.508475 | $\frac{15}{29}$ | 0.517241 |
| $\frac{19}{39}$ | 0.487179 | $\frac{43}{87}$ | 0.494253 | $\frac{29}{57}$ | 0.508772 | $\frac{44}{85}$ | 0.517647 |
| $\frac{39}{80}$ | 0.487500 | $\frac{44}{89}$ | 0.494382 | $\frac{28}{55}$ | 0.509091 | $\frac{29}{56}$ | 0.517857 |
| $\frac{20}{41}$ | 0.487805 | $\frac{45}{91}$ | 0.494505 | $\frac{27}{53}$ | 0.509434 | $\frac{43}{83}$ | 0.518072 |
| $\frac{41}{84}$ | 0.488095 | $\frac{46}{93}$ | 0.494624 | $\frac{26}{51}$ | 0.509804 | $\frac{14}{27}$ | 0.518519 |
| $\frac{21}{43}$ | 0.488372 | $\frac{47}{95}$ | 0.494737 | $\frac{25}{49}$ | 0.510204 | $\frac{41}{79}$ | 0.518987 |
| $\frac{43}{88}$ | 0.488636 | $\frac{48}{97}$ | 0.494845 | $\frac{49}{96}$ | 0.510417 | $\frac{27}{52}$ | 0.519231 |
| $\frac{22}{45}$ | 0.488889 | $\frac{49}{99}$ | 0.494949 | $\frac{24}{47}$ | 0.510638 | $\frac{40}{77}$ | 0.519481 |

to an approximating irreducible fraction one finds the bounding decimal values and picks one of the associated irreducible fractions.

One has to stand in awe of the amount of hand calculation that everyone including professional mathematicians performed in those days. But it was exactly the mind-numbing arduousness of these calculations, together with the

ever-present likelihood of error that that made numerical tables so highly valued
and the people who created accurate tables, such as Charles Haros, so greatly
appreciated. Perhaps it is not all that surprising that the first killer application
for the personal computer, Visicalc, was a table building program [**200**].

Haros started by generating the following sequence of fractions as the basis
of his induction:

$$f_i = \begin{cases} \frac{1}{99} & i = 1 \\ f_{i-1} + \frac{1}{a(a+1)} & f_{i-1} = \frac{1}{a+1} \quad \text{or} \quad f_{i-1} = \frac{a-1}{a} \end{cases}$$

He observed that these fractions are irreducible and that they have property
that $bc - ad = 1$ when $\frac{a}{b}$ and $\frac{c}{d}$ are adjacent fractions in the sequence.

He then set out to insert a fraction $\frac{x}{y}$ between any two elements $\frac{a}{b}$ and $\frac{c}{d}$
of this sequence that preserves this adjacency property. Solving for $x$ and $y$ he
found

$$\frac{x}{y} = \frac{a+c}{b+d}$$

Since $bc - ad = 1$, $\frac{a+c}{b+d}$ is irreducible but Haros only made the insertion if the
denominator, $b + d$, was less than or equal to 99. The fraction $\frac{x}{y}$ that Haros
solved for is of course the mediant of $\frac{a}{b}$ and $\frac{c}{d}$.

By means of demonstrations of the application of the algorithm, Haros
concluded that all the irreducible fractions between 0 and 1 with one or two
digit denominators were generated by his algorithm. A more mathematical
proof of this property of the Haros mediant algorithm would be supplied by
Cauchy in 1816, fourteen years in the future, but Haros' demonstrations were
a respected and acknowledged method of establishing mathematical certainty
in 1802. Lagrange, one of Haros' colleagues, used this proof technique widely
in his papers.

## 2.6. "Instruction abrégée sur les nouvelles mesures ..."

The actual tables that Haros created using the methods he described in his
note in the *Journal de l'École Polytechnique* appeared a paper he published in

1801, "Instruction abrégée sur les nouvelles mesures, avec des tables de rapports et de réduction" [**125**].

This paper was submitted to the Mathematics Section of the Institut de France where it was turned over for review to two institute members, Gaspard de Prony and Adrien-Marie Legendre. De Prony and Legendre gave the paper a glowing review.

> Citizen Haros is one of the geometricians in the computation section of the Cadastre bureau, where he has given ample demonstration of his knowledge and his talents. The commissaries think that the work of this citizen, whom they have just noticed in the class, can be extremely useful for promoting the understanding of the new system of weights and measures, and for facilitating its use.

The third part of this tract was a table that aided in the conversion between anciennes and nouvelles measures. At the end of the work under the title "Supplément" there was a table for evaluating an ordinary fraction as a decimal.

> The third part contains tables for reducing any number expressed in the old units into the new units by simple addition, and vice versa, with applications related to each table.
>
> At the end of this work, under the heading of a supplement, one can find a table for converting an ordinary fraction into a decimal with any degree of approximation one wishes. The specific arrangement of this table is based upon the remarkable property that every fraction that cannot be expressed exactly in decimals gives rise to a period.

Figure 2.3 is the complete report by Legendre and de Prony to the Institute on "Instruction Abrégée ..." and Figure 2.4 is an advertisement for the commercial version of the tables.

Of Haros' conversion tables, Lacroix says in the Foreword of his *Traité Élémentaire d'Arithmétique*

## SÉANCE DU 21 VENTÔSE AN 9.

### 35

A laquelle ont assisté les C<sup>rs</sup> Jeaurat, Tenon, Huzard, Desmarest, Bossut, Lassus, Ventenat, Bory, Tessier, Lamarck, Sabatier, Guyton, Deyeux, Lefèvre-Gineau, La Billardière, Fourcroy, Lagrange, Lelièvre, Olivier, Haüy, Vauquelin, Prony, Delambre, Brisson, Laplace, Cels, Berthollet, Charles, Monge, Lalande, Messier, Lacepède, Legendre, Lacroix, Coulomb, Des Essartz, Desfontaines, Jussieu, Cuvier, Parmentier, Hallé, Pelletan, Portal, Périer.

Le Président ouvre la Séance à cinq heures et demie du soir.

Les C<sup>rs</sup> Prony et Legendre font le Rapport suivant sur le Mémoire du C<sup>n</sup> Haros, intitulé *Instruction abrégée sur les nouvelles mesures:*

« La Classe nous a chargés, le C<sup>n</sup> Legendre et moi, de lui rendre compte d'un ouvrage du C<sup>n</sup> Haros, intitulé *Instruction abrégée sur les nouvelles mesures etc.*.

« Cet ouvrage est divisé en trois parties. La première renferme des notions préliminaires sur le nouveau système des poids et mesures, avec l'origine et la va-leur de chaque espèce de mesures, un tableau qui contient les noms synonymes des nouvelles mesures, leurs subdivisions décimales, leurs rapports avec l'unité fondamentale appelée *mètre*, leurs valeurs en mesures anciennes, avec les noms de ces anciennes mesures, leurs subdivisions et leurs valeurs en mesures nouvelles. On trouve, à la fin de cette partie, une table de rapports très approchés des nouvelles mesures aux anciennes, exprimés en nombres entiers.

« La deuxième partie renferme les principes du calcul décimal, appliqué aux nouvelles mesures, dans les quatre premières règles de l'arithmétique.

« La troisième partie contient des tables pour réduire, par une simple addition, un nombre quelconque de mesures anciennes en mesures nouvelles et réciproquement, avec des applications relatives à chaque table.

« On trouve à la fin de cet ouvrage, sous le titre de supplément, une table pour évaluer une fraction ordinaire en décimale, avec tel degré d'approximation qu'on voudra. La disposition particulière de cette table est fondée sur cette propriété remarquable, que toute fraction qui ne peut être évaluée exactement en décimales donne lieu à une période.

« Le C<sup>n</sup> Haros est un des géomètres de la section des cultivateurs du Bureau du cadastre, où il a donné des preuves soutenues de science et de talent; les Commissaires pensent que l'ouvrage de ce citoyen dont ils viennent de rendre compte à la Classe, peut être fort utile pour propager la connoissance du nouveau système des poids et mesures et en faciliter l'usage. »

Signé à la minute: Legendre et Prony.

FIGURE 2.3. Report of Legendre and de Prony on Haros' Tables

The presentation of the new units was naturally placed after the calculation of the decimal; and at the end of the work, I put the tables calculated in the offices of the Cadastre by Citizen Haros, by means of which one could convert the old units into the new ones and vice versa.

INSTRUCTION ABRÉGÉE *sur les nouvelles Mesures* qui doivent être introduites dans toute la république, au premier vendémiaire an 10; avec des tables de rapports et de réductions ; par *C. H. Haros*, employé au cadastre. Vol. *in*-12. Prix, 1 fr. 50 cent. et 1 fr. 80 cent. franc de port. Paris, rue de Thionville, n°. 116, chez *Firmin Didot*, libr. pour les Mathématiques, l'Architecture, la Marine, et les Éditions stéréotypes.

Cet ouvrage est divisé en trois parties.

La première renferme des notions préliminaires sur le nouveau système des poids et mesures. — La deuxième renferme les principes du calcul décimal, appliqué aux nouvelles mesures, dans les quatre premières règles de l'Arithmétique.—La troisième contient des tables pour réduir par une simple addition un nombre quelconque de mesures anciennes en mesures nouvelles, et réciproquement, avec des applications relatives à chaque table.

On trouve à la fin cet ouvrage sous le titre de *Supplément*, une table pour évaluer une fraction ordinaire en décimale, avec tel degré d'approximation qu'on voudra.

Les citoyens le *Gendre* et *Prony*, chargés d'en rendre compte à l'Institut national, terminent ainsi leur rapport : « Le citoyen *Haros*, » est un des géomètres de la section des calculateurs du bureau du » Cadastre, où il a donné des preuves soutenues de sciences et des » talents. Les commissaires pensent que l'ouvrage de ce citoyen dont » ils viennent de rendre compte à la classe, peut être fort utile pour » propager la connoissance du nouveau système des poids et mesures, » et en faciliter l'usage. »

FIGURE 2.4. Advertisement for Haros' Tables

Haros' "Instruction abrégée ..." tract was described in a paper by Roger Mansuy in 2008 [171]. At the time Mansuy located a copy of the tract in the Bibliothèque nationale de France but the response to a recent inquiry to the BnF indicated that it could no longer be located. There is however a copy at the Bibliothèque des Sciences et de l'Industrie.

## 2.7. Computing Logarithms

While at the Bureau de Cadastre de Paris, Charles Haros worked for Gaspard Riche de Prony [23]. De Prony led the computation of the Great Logarithmic and Trigonometrical Tables computed by the Bureau at the end of the

$18^{th}$ century. [**116**]. Beside co-authoring a table of logarithms together with Plausol and Bauzon, [**129**] Haros is credited in the literature with two formulæ to help in computing and checking tables of logarithms. In [**89**] he is credited with the following formula:

$$\log(x+2) = \log(x-2) + 2\log(x+1) - 2\log(x-1)$$
$$- \frac{2}{la}\left[\frac{2}{x^3-3x} + \frac{1}{3}\left(\frac{2}{x^3-3x}\right) + \text{etc.}\right] \qquad (2.7.1)$$

In [**88**], [**160**] and [**232**] Haros is credited with the following formula:

$$\log(x+5) = \log(x+3) + \log(x-3) + \log(x+4) + \log(x-4)$$
$$- \log(x-5) - 2\log(x)$$
$$- \frac{2}{la}\left[\frac{72}{x^4-25x^2+72} + \frac{1}{3}\left(\frac{72}{x^4-25x^2+72}\right) + \text{etc.}\right] \qquad (2.7.2)$$

where $la$ is the logarithm to the base $e$ of the base to which the logarithm is being computed.

These two series are referenced by Lacroix in *Complément des Élémens d'Algèbre* [**150**] and were still being studied and improved as late as 1812 by Thomas Lavernéde. In his *Treatise on Algebra* [**20**] Bonnycastle refers to "the transformations of Borda and Haros which are so often cited by foreign writers on the subject."

Formulæ such as these were used to reduce the number of computations needed to derive a new entry in the table when $x$ was so large or small that all terms after the first few could be ignored. This simplicity may have been especially sought after at a time that the people doing the actual computation might well have been grumpy ex-hairdressers with fewer royal heads to dress thanks to the guillotine[**116**]. These formulæ were also useful to perform checks on completed computations as they encoded mathematical consistencies that necessarily existed among tabular values.

```
ApproximateLog[ x_, d_] := N[Log[x], d]

RationalLog[x_, dx_] := Rationalize[Log[x], dx]

HarosTwoStepLog[np2_, m_, Lf_, p_] := Module[{v, s, n},
```

```
n = np2 - 2;
s = Sum[(1/k) (2/(n^3 - 3 n))^k, {k, 1, 2 m - 1, 2}];
v = Lf[n - 2, p] + 2*Lf[ n + 1, p] - 2*Lf[n - 1, p];
v + 2 s
]

HarosFiveStepLog[np5_, m_, Lf_, p_] := Module[{v, s, n},
n = np5 - 5;
s = Sum[(1/k) (72/(n^4 - 25 n^2 + 72))^k, {k, 1, 2 m - 1, 2}];
v = Lf[n + 3, p] + Lf[n - 3, p] + Lf[n + 4, p] + Lf[n - 4, p] -
Lf[n - 5, p] - 2 Lf[n, p];
v - 2 s
]
```

However, such formulæ do have the unfortunate downside that any error in a single value could be spread among many others. It was the use of such formulæ, coupled with the employment of mathematical innocents to perform the computations that led Edward Sang to opine, "The method followed in the calculation of the Cadastre table of logarithms was an egregious blunder. The result was in accordance with the method." [213]. Of course Sang was at the same time pitching a project to compute his own tables [212].

The following table that appeared in a paper by Thomas Lavernéde in the *Annales de Mathématiques Pures et Appliquées* [160] nicely summarized the various logarithm formulæ that were in use in 1810. The numbers that are part of the description refer to equations described in the paper.

TABLE 2.8. Accuracy of Approximation Equations for Logarithms

| Formula | $10^2$ | $10^3$ | $10^4$ | $10^5$ | $10^6$ | $10^2$ | $10^3$ | $10^4$ | $10^5$ | $10^6$ |
|---|---|---|---|---|---|---|---|---|---|---|
| $2^{nd}$ Degree, #39 | 4 | 6 | 8 | 10 | 12 | 13 | 19 | 25 | 31 | 37 |
| M. de Borda #40 | 5 | 8 | 11 | 14 | 17 | 17 | 26 | 35 | 44 | 53 |
| M. Haros #41 | 6 | 10 | 14 | 18 | 22 | 18 | 30 | 42 | 54 | 66 |
| $4^{th}$ Degree #42 | 6 | 10 | 14 | 18 | 22 | 20 | 32 | 44 | 56 | 68 |
| $5^{th}$ Degree #44 | 6 | 11 | 16 | 21 | 26 | 19 | 34 | 49 | 64 | 79 |
| $5^{th}$ Degree #45 | 6 | 11 | 16 | 21 | 26 | 20 | 35 | 50 | 65 | 80 |
| $6^{th}$ Degree #46 | 8 | 14 | 20 | 26 | 32 | 24 | 42 | 60 | 78 | 96 |

## 2.8. General Purpose Root Finder

Haros is also cited by Clairaut in *Éléments d'Algèbre* ([42], p.211) as well as in Hutton's Tracts on Mathematical and Philosophical Subjects [138] for independently discovering and then improving on a formula of Lambert for computing roots. Clairaut says

> This is where the formula was when Haros, a mathematician in the land registry, not only found it without knowing of the work of Lambert but perfected it ...

In a review of the 1812 edition of *Tracts on Mathematical and Philosophical Subjects* by Charles Hutton [138] that appeared in the July, 1813, issue of The Quarterly Review, the reviewer offered the following comment about Tract 10:

> Tract 10 contains the investigation of some easy and general rules for extracting any root of a given number. ... The only rule which has ever been put in competition with this is that of M. Haros, which is
>
> $$\sqrt[n]{a^n \pm d} = a \pm \frac{2ad}{2na^n \pm (n-1)d},$$
>
> where $N = a^n \pm d$, or $d =$ the difference between the assumed power and the given number. It is not a little extraordinary that the English admirers of M. Haros' formula should not

have discovered that it is no other than the rational formula of Halley published in 1694.

```
HarosRootFinder[v_, n_, e_] :=
 Module[{a, d, r = 1, k = 0},
  While[r^n < v, r++];
  a = Which[v - (r - 1)^n < r^n - v, r - 1, True, r];
  While[Abs[a^n - v] > e] ,
   k++;
   d = v - a^n;
   a = a + 2 a d / (2  n a^n  + Sign[d]* (n - 1) d);
   ];
  {v, a, a^n, a^n - v, k}
 ]
```

## 2.9. Haros' Publications

Today we would regard Charles Haros as a computer scientist. From what little we can glean about him he comes through as being very much of the same bent as the great British table maker, E.H. Neville, whom we will meet shortly.

Like Neville, Haros was a table maker and also in that context, like Neville, the inventor of algorithms, formulæ and identities that aided computation. Table 2.9 is a partial list of Haros' published work. There were at least three editions of *Instruction abrégée aux nouvelles mesures*. Only the first and the third are cited.

Figure 2.5 is a page from Lacroix's textbook on arithmetic [151] that demonstrates the use of Haros' 1806 tables for converting measures.

Haros also contributed tables to books authored by others. Figure 2.6, for example is taken from Ramel's tract *Système Métrique, où Instruction Abrégée sur les Nouvelles Mesures* published in 1808 [204]. In the introduction Ramel credits the help of Haros so it is quite possible that the fraction-to-decimal tables are exactly those in Haros' tract. Table 2.9 lists this contribution along

D' A R I T H M É T I Q U E.    147

*Tables calculées par M. Haros, pour la conversion des mesures anciennes en nouvelles, et réciproquement.*

EXEMPLES DE L'USAGE DE CES TABLES.

Chaque colonne contient, en première ligne, la valeur de la mesure désignée dans le titre de la colonne, et ensuite les produits de cette valeur par les nombres écrits dans la colonne marquée *N*.

Il y a partout cinq décimales, mais dans l'usage ordinaire on peut se borner à trois.

Convertir 8 toises 5 pieds 7 pouces, en mètres.

| | |
|---|---|
| 8$^t$ valent........ | 15$^m$,592 |
| 5$^{pi}$ ............. | 1 ,624 |
| 7$^{po}$ ............. | 0 ,189 |
| Somme..... | 17$^m$,405 |

Rép. 17 mètres 40 centimètres.

Convertir 89 aunes ⅞ de Paris, en mètres.

| | |
|---|---|
| 80$^{aun.}$ valent..... | 95$^m$,976 |
| 9 ............. | 10 ,696 |
| ⅞ ............. | 0 ,891 |
| Somme........ | 106$^m$,663 |

Rép. 106 mètres 66 centimètres.

Convertir 218 arpens ( eaux et forêts ) en hectares.

| | |
|---|---|
| 200$^{arp.}$ valent | 102$^{hect.}$,144 |
| 10 ........ | 5 ,107 |
| 8 ........ | 4 ,086 |
| Somme ... | 111$^{hect.}$,337 |

Rép. 111 hect. 34 ares environ.

Convertir 3050 livres (de poids) en kilogrammes.

| | |
|---|---|
| 3000$^{liv.}$ valent | 1468$^{kilos.}$,52 |
| 50 ........ | 24 ,48 |
| Somme...... | 1493$^{kilos.}$,00 |

Rép. 1493 kilogrammes.

Les mêmes tables peuvent donner le prix de la nouvelle unité d'une matière par celui de l'ancienne unité, exprimé en francs. Par exemple, l'aune de drap coûtant 37$^{fr}$,50$^c$, il est visible que si l'on connaissait l'expression du mètre en aune, il n'y aurait qu'à multiplier cette expression par 37,5 ce qui reviendrait à convertir 37$^m$,5 en aunes et parties décimales de l'aune ; mais il faudrait compter le résultat pour des francs. Voici le calcul de cet exemple.

Par la table qui convertit les mètres en aunes :

| | |
|---|---|
| 30$^m$ valent............... | 25$^{aun.}$,243 |
| 7 ............. | 5 ,890 |
| 0 ,5 ............. | 0 ,421 |
| 37$^m$,5 ............. | 31$^{aun.}$,554 |

et prenant ce résultat pour des francs, on trouve 31 fr. 55 cent. pour le prix du mètre de drap.

Lorsqu'on veut convertir les nouvelles mesures dans les anciennes, on n'obtient, par les tables suivantes, que des entiers et des fractions décimales, et il reste à convertir ces fractions en subdivisions propres à chaque espèce de mesure.

2

FIGURE 2.5.  Use of Haros' 1806 Measure Conversion Tables

TABLE 2.9. Books and Papers (Co-)Authored by Charles Haros

| Year | Title |
|------|-------|
| 1801 | Instruction abrégée aux nouvelles mesures [125],[126] |
| 1802 | Tables pour évaluer une fraction ordinaire ... [128] |
| 1802 | Comptes faits à la maniére de Barême ... [127] |
| 1803 | Traité de l'arpentage et du toisé ... [190] |
| 1805 | Tables de logarithme à l'usage des ingénieurs du cadastre ... [131] |
| 1806 | Tables de logarithme à l'usage des ingénieurs du cadastre ... [129] |
| 1806 | Tables de Logarithmes et des tables pour la conversion des nouveaux poids et mesures [130] |

TABLE 2.10. Haros' Contributed Tables

| Year | Author | Title |
|------|--------|-------|
| 1804 | Sylvestre François Lacroix | Traité Élémentaaire d'Arithmétique |
| 1806 | L'Abbe Delagrive | Manuel de Trigonométrie Pratique |
| 1808 | Ab. Ls. Ramel | Système Métrique, oú Instruction Abrégée sur les Nouvelles Mesures |
| 1811 | Étienne Bezout | Traite D'Arithmétique |
| 1821 | Cyprien Prosper Brard | Mineralogie Appliquee aux Arts |

with other books to which Haros, singly or in collaboration with other members of the bureau, provided tables.

At the same time he was building tables for metric conversion and tables of logarithms, Haros was deeply involved in computing the famous French ephemeris, *Connaissance des Temps ou des Mouvemens Célestes, à l'Usage des Astronomes et des Navigateurs*. A review of the ephemeris for year 1810 (published in 1808) in *The Eclectic Review for August 1810* states

> The volume (of the *Connaissance des Temps*) for 1795 was published by the temporary commission of weights and measures. From that time the calculations of this Ephemeris have been usually made by M.M. Haros and Marion, under the inspection and direction of the Board of Longitude ...

| 17 | 0,7.$\frac{8}{11}$ | | 17 | 0,7083 etc. | | 17 | 0,6.$\frac{7}{13}$ |
|----|----|---|----|----|---|----|----|
| 19 | 0,8.$\frac{7}{11}$ | | 19 | 0,7916 etc. | | 19 | 0,7.$\frac{4}{13}$ |
| 21 | 0,9.$\frac{6}{11}$ | | 23 | 0,9583 etc. | | 21 | 0,8 $\frac{6}{13}$ |
| | | | | | | 23 | 0,8.$\frac{11}{13}$ |
| | | | | | | 25 | 0,9.$\frac{8}{13}$ |

**23.**   **25.**   **27.**

| 23. | | | 25. | | | 27. | |
|----|----|---|----|----|---|----|----|
| 1 | 0,0 | | 1 | 0,04 ex. | | | |
| 10 | 0,4 | | 2 | 0,08 ex. | | 1 | 0,0 |
| 8 | 0,3 | | 3 | 0,12 ex. | | 10 | 0,3 |
| 11 | 0,4 | | 4 | 0,16 ex. | | 19 | 0,7 pér. |
| 18 | 0,7 | | 6 | 0,24 ex. | | | |
| 19 | 0,8 | | 7 | 0,28 ex. | | 2 | 0,0 |
| 6 | 0,2 | | 8 | 0,32 ex. | | 20 | 0,7 |
| 14 | 0,6 | | 9 | 0,36 ex. | | 11 | 0,4 pér. |
| 2 | 0,0 | | 11 | 0,44 ex. | | | |
| 20 | 0,8 | | 12 | 0,48 ex. | | 4 | 0,1 |
| 16 | 0,6 | | 13 | 0,52 ex. | | 13 | 0,4 |
| 22 | 0,9 | | 14 | 0,56 ex. | | 22 | 0,8 pér. |
| 13 | 0,5 | | 16 | 0,64 ex. | | | |
| 15 | 0,6 | | 17 | 0,68 ex. | | 5 | 0,1 |
| 12 | 0,5 | | 18 | 0,72 ex. | | 23 | 0,8 |
| 5 | 0,2 | | 19 | 0,76 ex. | | 14 | 0,5 pér. |
| 4 | 0,1 | | 21 | 0,84 ex. | | | |
| 17 | 0,7 | | 22 | 0,88 ex. | | 7 | 0,2 |
| 9 | 0,3 | | 23 | 0,92 ex. | | 16 | 0.5 |
| 21 | 0,9 | | 24 | 0,96 ex. | | 25 | 0,9 pér. |
| 3 | 0,1 | | | | | | |
| 7 | 0,3 pér. | | **26.** | | | 8 | 0,2 |
| | | | | | | 26 | 0,9 |
| | | | | | | 17 | 0,6 pér. |

**24.**   **26.**   **28.**

| 24. | | | 26. | | | 28. | |
|----|----|---|----|----|---|----|----|
| | | | 1 | 0,0.$\frac{5}{13}$ | | | |
| | | | 3 | 0,1.$\frac{2}{13}$ | | | |
| 1 | 0,0416 etc. | | 5 | 0,1.$\frac{12}{13}$ | | 1 | 0,03.$\frac{4}{7}$ |
| 5 | 0,2083 etc. | | 7 | 0,2.$\frac{9}{13}$ | | 3 | 0,10.$\frac{5}{7}$ |
| 7 | 0,2916 etc. | | 9 | 0,3.$\frac{6}{13}$ | | 5 | 0,17.$\frac{6}{7}$ |
| 11 | 0,4583 etc. | | 11 | 0,4.$\frac{3}{13}$ | | 9 | 0,32.$\frac{1}{7}$ |
| 13 | 0,5416 etc. | | 15 | 0,5.$\frac{10}{13}$ | | 11 | 0,39.$\frac{2}{7}$ |

FIGURE 2.6. Table Possibly from Instruction Abrégée

The foreword to that edition of the ephemeris describes Haros' extensive involvement in this effort:

> The calculations were made, as was usual, under the surveillance of the Bureau des Longitudes, by MM. Haros and Marion, using M. Burg's tables for the moon, those of the late M. Lalande for Mercury, Venus and Mars, and using those of M. Delambre for the sun, Jupiter, Saturn, Uranus and the satellites. The lunar and solar eclipses and the occultations of stars were calculated by M. Matthieu, Assistant Secretary of the Imperial Observatory.

## 2.10. The Bureau du Cadastre

Over the opposition of Citizen Cambon, the creation of the Bureau du Cadastre was finally approved on March 22, 1793. [**47**] de Prony who had been the provisional director since September 23, 1790, when the bureau was initially decreed, became its director. de Prony was also the head of *Ponts & Chaussées* at the time.

The Bureau du Cadastre was part of the Bureau des Longitudes where Lagrange and Laplace were in charge of surveying and Lalande, Missier, Méchain and Delambre were in charge of astronomy.

The charter of the Bureau du Cadastre was as follows:

> Description géographique & topographique du territoire de la Reépublique; Formation de diverses Cartes pour la Géographie Physique & Économique; Calcul des grandes Tables Logarithmiques, d'après le nouveau Systême des Poids & Mesures; Calcul de la Connoissance des Temps, &c.
>
> Les géographes employés au cadastre sont pris dans l'école instituée par le décret du 30 vendémiaire.
>
> *Voyez* l'article de cette Ecole, faisant partie de celui des Ecoles de services publics.

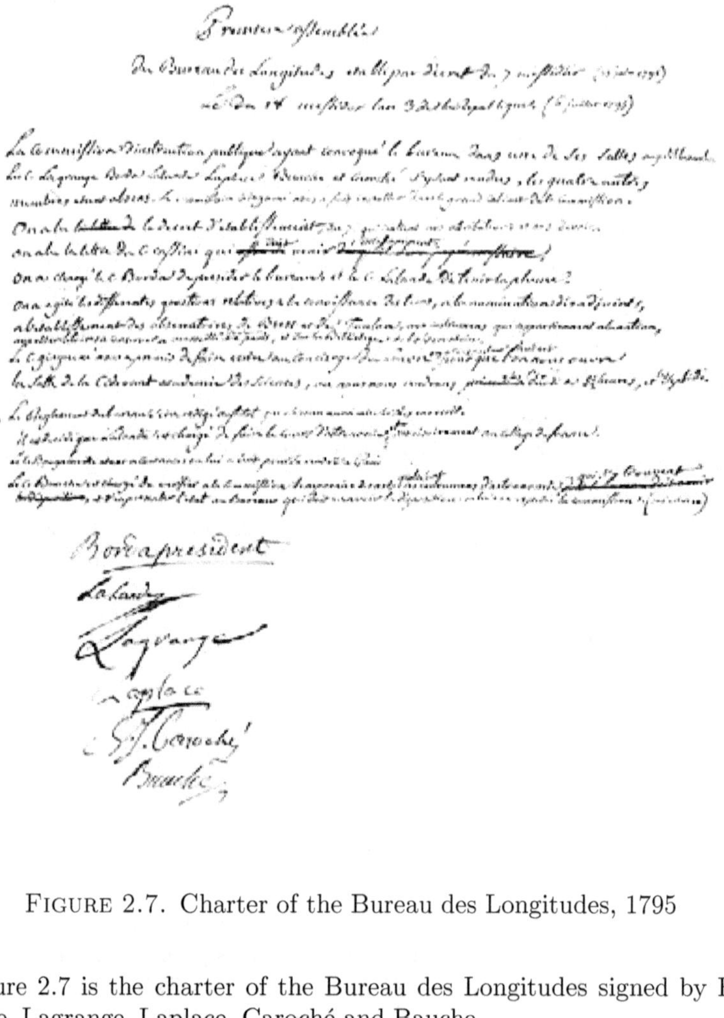

FIGURE 2.7. Charter of the Bureau des Longitudes, 1795

Figure 2.7 is the charter of the Bureau des Longitudes signed by Borda, LaLande, Lagrange, Laplace, Caroché and Bauche.

From its founding until 1802 the Bureau du Cadastre was part of the Bureau des Longitudes. In 1802 Bureau du Cadastre was folded into École des Ponts et Chaussées along with the École des Géographes. The bureau's first, last and only Directeur, de Prony, was also the Directeur of École des Ponts et Chaussées and at this time became a *Membre Surnuméraire* of the Bureau des Longitudes so the association of the Bureau du Cadastre mathematicians was still in place even if the bureau itself had ceased to exist. In fact these mathematicians were still being referred to as "employés au Cadastre" as well "attachés au Cadastre", "ingénieurs des Ponts et Chaussées", "professeurs á École des Ponts et Chaussées" as late as 1806 [**197**].

At the time Haros wrote his 1802 paper, when Bureau du Cadastre was still in existence his reporting structure looked like this:

Napoleon Bonaparte, Premier Consul

   Jean-Antoine Chaptal, Ministre de l'Intérieur

      Antoine-Vincent Arnault, Chef, Instruction publique

         Amauri-Duval, Chef, Suite du Buréau des Beaux-Arts

            Louis Antoine de Bougainville (?), Chef, Bureau des Longitudes

               Gaspard Marie Riche de Prony, Directeur, Bureau du Cadastre

                  Jean-Jacques Lequeu, Dessinateur du Section des Calculateurs

                     Charles Haros, Géomètre au Cadastre

Bougainville is occasionally referred to as the chief of the Bureau des Longitudes but there is no chief listed in the *Almanach National de France of 1802*. The Bureau des Longitudes was allowed to choose its own chief by unanimous vote of the board members so perhaps this is the reason that nothing was recorded in the Almanach. Bougainville was one of two "anciens navigateurs." The other one was Borda.

I have not found a street address for the Bureau du Cadastre. It may have been in the Ministry of the Interior on rue de Grenelle, just east of the Hotel Royal des Invalides or it may have been located somewhere down south near the Observatory. Figures 2.8 and 2.9 are clipped from a map drawn by W.B.

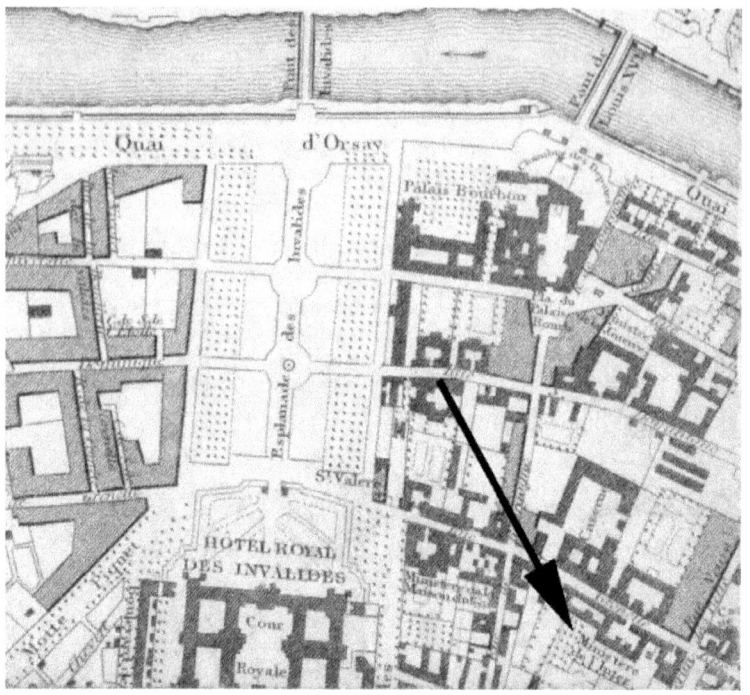

FIGURE 2.8. Location of the Ministry of the Interior in 1833

Clarke and published by the Society for the Diffusion of Useful Knowledge in 1833. The Ministry of the Interior can be seen in the lower right-hand corner of the map in Figure 2.8. Figure 2.9 locates the Observatory in 1833. It is roughly where the Bureau des Longitudes is today except that Rue d'Enfer has become Avenue Denfert-Rochereau.

The historical record is silent as to whether Napoleon ever stopped to take the guys in the Bureau out for a Friday afternoon beer. Maybe he just didn't want to run into Laplace again. The following are excerpted from the Gleanings Far and Near section of *The Mathematical Gazette*, Vol. 15, No 211 (January, 1931) and Vol. 19, No. 235 (October, 1935) respectively:

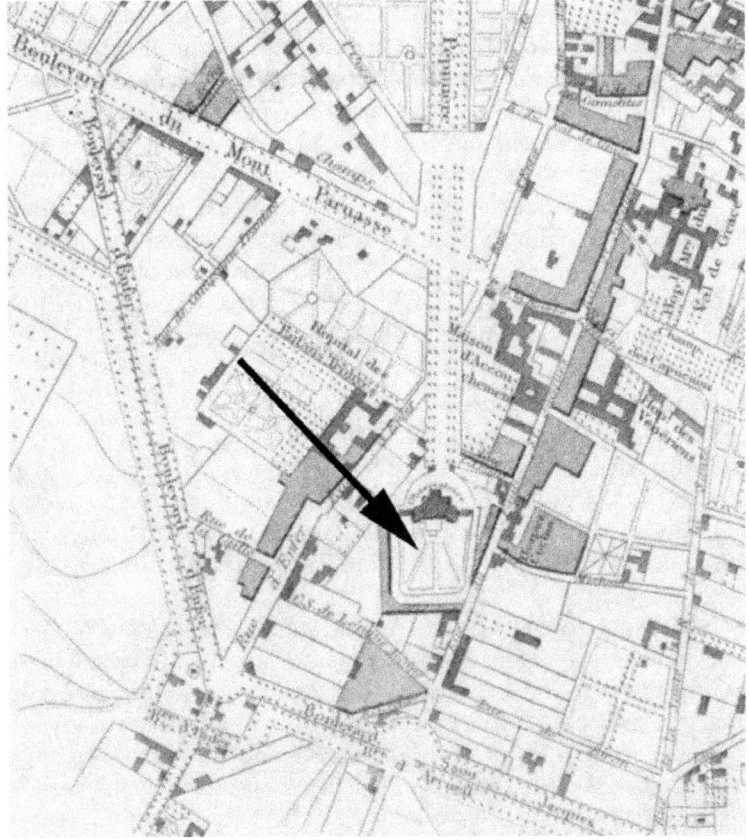

FIGURE 2.9. Location of the Paris Observatory in 1833

The celebrated Laplace, then Examiner in the artillery school
[at Chalons], was a man of the most serious appearance: his sad
and severe face, his black dress, his fringed ruffles, the shade
over his eyes – rendered necessary by the state of his sight –
gave him a very imposing air . . . It may be imagined with what
anxiety, disquietude, and sinking of the heart we approached
the Examiner's table . . .

NAPOLEON EXAMINED BY LAPLACE. We have no record
of the questions or of the effect that Napoleon made upon the
man who was the greatest in Europe at his trade, but we know
something of the scene: the large hall in the École Militaire
with its ornamented curtained windows, the two big slates set
up for the chalk diagrams of the students, the benches rising
in tiers so that all the students would be seen at once by the
examiner, and Laplace in his black broadcloth sitting there
with a shade over his weak eyes, asking his questions in turn.
Four gentlemen cadets passed out from that examination to
their commission, Napoleon third on the list.

The members of the Bureau du Cadastre including Haros were actually
employees of the Ecole des Géographes du Cadastre so this may be why Haros
is occasionally referred to as a Professeur. At the time de Prony was also *chef
des Ponts & Chaussées*.   Wherever the nature of de Prony's little group was
we can be sure that it was right in the middle of the construction of the Tables
du Cadastre for taxation and the Grandes Tables of logarithms.

On the title page of de Lagrive's *Manuel de Trigonométrie Pratique* of 1806
[**61**] Haros is mentioned along with Reynaud, Plausol and Bauzon as being an
attaché au Cadastre. Haros contributed a number of tables to this book for
converting between old and new measures. In the publisher's advertisement
for this book Reynaud, Haros, Plausol and Bauzon are referred to as "les Professeurs du Cadastre." Haros also contributed tables to *Arithmétique de Bezout*
by Reynaud [**207**].

Table 2.11 is a list of some of the people that Haros worked with in the
Bureau des Longitudes and the Bureau du Cadastre according to [**188**] among
others.

TABLE 2.11.   A Sampling of Charles Haros' Colleagues

| Name | Title |
|------|-------|
| Arago, François | Membre du Bureau des Longitude |
| Bauche, M. | Géographe au Bureau des Longitudes |
| Bauzon, M. | Membre du Bureau du Cadastre |
| Biot, Jean-Baptiste | Membre du Bureau des Longitude |
| Bougainville, Louis Antoine de | Anciens Navigateur |
| Calmels, L. | Géomètre en Chef du Cadastre |
| Caroché, M. | Artiste au Bureau des Longitudes |
| Borda, Jean-Charles | Inspecteur des Constructions |
| Cousin, Jacques A. J. | Professeur au Collége de France |
| Delambre, Jean Baptiste Joseph | Astronome du Bureau des Longitudes |
| de Prony, Gaspard Marie Riche | Chef des Ponts & Chaussées |
| Fleurieu, Charles Pierre Claret | Anciens Navigateur |
| Garnier, Jean-Guillaume | Chef du Bureau du Cadastre |
| Hennet, A.M. | Commissaire Impérial du Cadastre |
| Legendre, Adrien-Marie | Membre Bureau des Longitude |
| Lagrange, Joseph Louis | Géomètre au Bureau des Longitudes |
| Lagrive, Jean de | Membre du Bureau du Cadastre |
| Lalande, Joseph Jérôme | Astronome au Bureau des Longitudes |
| Laplace, Pierre-Simon | Géomètre au Bureau des Longitudes |
| Lefévre-Gineau, Louis | Ingénieur en Chef du Cadastre |
| Mauduit, Antoine-René | Lecteur Royal en Mathématiques |
| Méchain, Francois André | Astronome au Bureau des Longitudes |
| Messier, Charles | Astronome dans Bureau des Longitude |
| Missier, M. | Astronome au Bureau des Longitudes |
| Oyon, J.B. | Chef du Bureau du Cadastre |
| Parseval, Marc-Antoine | Membre du Bureau du Cadastre |
| Paucton, Alexis-Jean-Pierre | Employé au Bureau du Cadastre |
| Peuchet, Jacques | Ministére de L'intérieur |
| Plauzoles, Charles de | Chef du Bureau du Cadastre |
| Plausol, M. | Membre du Bureau du Cadastre |
| Reynaud, Antoine-Andre-Louis | Professeur des Élèves du Cadastre |
| Salverte, Eusebe-Baconniere de | Membre du Bureau du Cadastre |

## 2.11.  Grandes Tables du Cadastre

The primary activity of the Bureau du Cadastre in its early years was the computation of de Prony's *Grandes Tables du Cadastre* that included logarithms to twenty-six places. Workers on this project were divided into three tiers. At the top were five or six mathematicians that developed the mathematical formulæ to be used. The second tier consisted of seven or eight "calculators" who turned the formulæ produced by the first tier into step-by-step instructions for the seventy or eighty members of the third tier that performed the actual computations.

The computational plans prepared by the second tier calculators were assembly language programs for human dataflow computers [8]. Each piece of paper passed from one hairdresser to the next was a token that held the state of a computation and each hairdresser was an order code.    In our effort to bring Charles Haros into high relief, we need to explore whether he was in the first or second tier in de Prony's human computer.

In his detailed and exhaustive review of the construction of the Tables du Cadastre [163], Lefort describes the first two tiers of de Prony's calculating organization as follows:

> La première section, composée de cinq ou six géomètres, s'occupait de la partie purement analytique, et du calcul de quelques nombres fondamentaux.
> La deuxième section contenait sept ou huit calculateurs, possédant l'analyse et ayant une grande pratique de la traduction des formules en nombres.

Lefort puts de Prony and Legendre in the first tier and Grattan-Guinness adds Prieur and Carnot [115]. With respect to the second tier, Lefort contributes Letellier and Guyétant and Grattan-Guinness adds Parseval and Garnier.

As we have seen, Haros produced original and respected mathematics for computing logarithms which would make him a candidate for membership in the first tier. Furthermore, he is not infrequently referred to as a *géomètre du cadastre.* For example, in his "Suite des Leçons d'Analyse" in 1796 de Prony says

> *Lans* et *Haros*, deux géomtres du cadastre, se sont occupés avec
> beaucoup de succès du problème de *Mouton* ... [**56**]

On the other hand, it must be said that his talents, at least as we know them, fit the description of members of the second tier quite a bit better than the first. Indeed, it could be said they fit the description of the second tier perfectly. Furthermore, as it was the responsibility of the second tier to check the computations of the third tier it could be argued that Haros' logarithm formulæ are more appropriate for this than for generating the logarithms in the first place.

Next we note that Garnier refers to Haros as "l'un de mes collégues à l'ancien Bureau du Cadastre." [**88**] It's doubtful that Garnier would have referred to Legendre in this familiar manner.

Finally according to Grattan-Guinness [**114**], "When the project was completed, some of the calculators were transferred to the *Bureau des Longitudes* to work on astronomical tables..." and we have already cited a report that this is where Haros was by 1810.

Based on this, I'd put Charles Haros in de Prony's second tier, the calculators, but add quickly that only at this time and in this place would being regarded as a peer of mathematicians of the stature of Jean Garnier and Antoine Parseval be regarded as second tier.

## 2.12. Sources of Inspiration

Going forward from Haros, there are dots that we can sometimes only connect with conjectures about the history of a particular sequence of vulgar fractions together with its properties. Prior to Haros we can find the sequence and we can find the mediant property but not the two explicitly linked to each other. We are thus led to inquire about Haros' possible source of inspiration to unite the two.

We can be certain that Haros knew about the mediant, for example, as a method to find roots of equations, whether or not he knew specifically of the work Nicolas Chuquet or Estienne de la Roche. His root finding algorithm shows that it is likely that he was familiar with the work of both Edmond Halley and Gerardus Mercador.

There was an abundance of computation with continued fractions during this period so he certainly knew of the work of one of his mentors, Joseph Louis Lagrange. There is a tantalizing hint in the *Métrologies Constitutionnelle et Primitive* [**2**] published in 1801. It's hard to imagine that Haros didn't have a hand in producing this document but the only place where he is mentioned is in Article X entitled "On the use of continued fractions for the determination and confirmation of the ratios described above, and others, more or less closely approximated" where it is stated:

> The same method has just been used recently by the author of the abridged guide to the new units (C.H. Haros, geometrician employed at the Cadastre), which were approved on the 21st of Ventose in year 9 by the National Institute; in effect, the author announces there that it is on the basis of the theory of continued fractions that he is providing the relationships of the new units to the old ones. They are expressed in whole numbers, like those above, but more simply; in addition, they are less closely approximated, and yet of a precision that is sufficient for the needs of commerce, general practice, for very small weights, such as for grain, as well as for medium weights.

This passage refers to Haros' paper *Instruction abrégée aux nouvelles mesures* ... not *Tables pour évaluer une fraction ordinaire* ... but it does demonstrate that Haros was using continued fractions to create tables.

Finally, recall that logarithms are proportions and that the mediant is the proper way to combine proportions. Haros' organization and his logarithm formulæ tell us that he was deeply invested in building tables, especially tables of logarithms.

August Aubry published a paper in 1906 that bears directly on the history
of the Farey sequence before Haros. It was entitled "Les Logarithmes Avant
Neper" [**9**]. In a footnote in this paper, Aubry says:

> Estienne de la Roche, in his (Lyon, 1520) Arithmetique also
> used this algorithm, where it is called *médiacion*.
>
> The type of procedure just mentioned is today referred to
> as *mediants*, and their study has led to a number of different
> series that were studied by Farey, Cauchy, Brocot, Halphen,
> etc.

The procedure Aubry is talking about is Chuquet's *règle des nombres moyens* so
we have a solid connection between the computation of tables of logarithms and
the mediant. While Aubry fails to directly connect the rule to Haros he does
draw the logarithm line from Pythagoras through Chuquet to Farey. Haros was
thoroughly familiar with the use of continued fractions and the mediant and was
probably using one or the other to build tables of logarithms. Using these tools
to build his conversion tables would have been just as natural as documenting
how the mediant provided assurance that he found all $3,003$ vulgar fractions in
his 1802 paper describing the tables. This would give a mathematical heft to
the paper beyond simply describing the tables and their use.

## 2.13. Bookends on the Era of Organized Scientific Computation

Charles Haros was to Gaspard de Prony in the French *Tables du Cadastre*
project what Gertrude Blanch was to Arnold Lowan in the American Mathematical Tables Project. Both Haros and Blanch reported to the director of the
table project and both built programs for the human computers that performed
the actual computations.

Haros and Blanch, just as much as de Prony and Lowan, are bookends on
the period that Grier referred to as the era of organized scientific computation [**119**]. Charles Haros' work on the *Connaissance des Temps* contributed
to Charles Babbage's vision that computations should be done with steam.
Gertrude Blanch was ultimately displaced by the operating system in an electronic scientific computer, the IBM 1620.

## 2.14. Henry Goodwyn, Brewer and Table Maker

An obituary for Henry Goodwyn appeared in the *Mechanics' Magazine Museum, Register, Journal, and Gazette* of Saturday, April 19, 1828 [3]. The article is signed simply "The Mechanics' Magazine's Old and Cordial Friend." Based on the level of familiarity with Goodwyn and the writing style I suspect the author is Olinthus Gregory but this is pure conjecture. Here are a couple of paragraphs from the obituary describing Henry Goodwyn:

> Mr. Goodwyn, who was for some years the principal of the firm of Goodwyn and Co., brewers, was obliged to retire from the pursuits of active life nearly thirty years ago, in consequence of severe lameness and other disorders. Happily, however, he slid into retirement with competent property; and, being surrounded by an affectionate family, whose constant aim was to promote his comfort, he passed not merely with content, but with delight, from the occupations of a commercial man to those of a student, and at length to those of a student, and at length to those of a most laborious computer.
>
> Mr. Goodwyn being not a man of profound science, had too much good sense to attempt higher inquiries, but devoted himself to the investigation of the properties of numbers, among which those which relate to circulates engaged most of his time and thoughts. Being often confined to his bed for weeks, nay months, by his complicated disorders, he invented various mechanical expedients for the abridgement, as well as the correctness, of his computations, and frequently was he enables to derive such pleasure from his results (new to himself, and, in many cases, new to the world), as completely to overcome the sense of pain.

As a younger man, Henry Goodwyn owned and ran the Red Lion Brewery in the St. Katharine's section of London ([172], [136]). Even during this period of his life he was an avid and inveterate table builder producing all sorts of tables that had to do with brewing and distributing his products [98]. What led him to take up building a table of fractions and decimal equivalents in his retirement is a mystery but certainly the study of circulating decimals was a popular pastime among mathematical amateurs and savants alike.

According to *The Environs of London* of 1796 [**169**], a Henry Goodwyn, Esq., lived in what was then called Bastile House and is now called Vanbrugh (also Vanburgh) Castle after its architect, Sir John Vanbrugh. Figure 2.10 is an approximate location of Maze Hill and Bastile House on a clip from the map of Kent and Essex in *Environs*. It is worth noting that Bastile House is between the Greenwich Observatory, a bastion of table production, and the Royal Military Academy at Woolwich where Dr. Olinthus Gregory. a close friend of Goodwyn's and another amateur number theorist, taught.

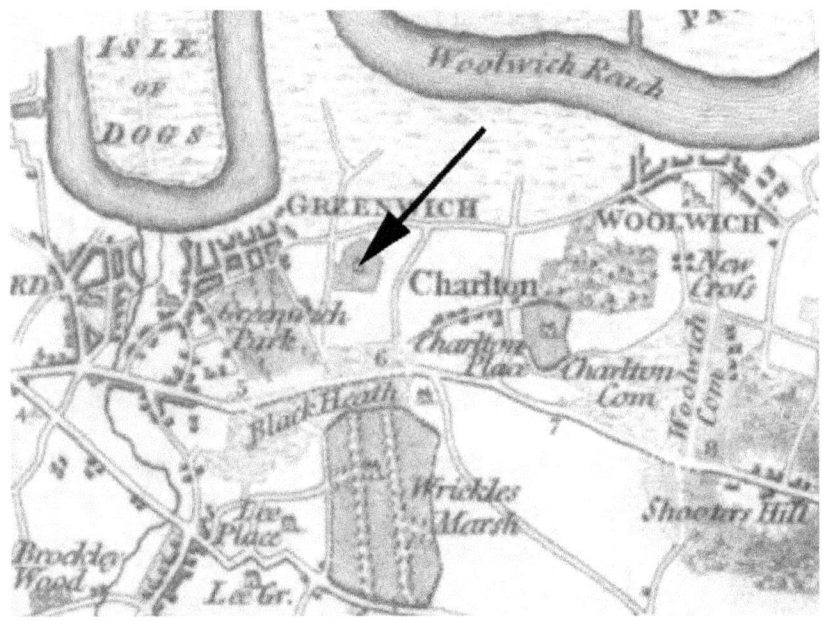

FIGURE 2.10.   Location of Maze Hill and Bastile House

The journey of the mediant property of the sequence of vulgar fractions with denominators less that a given value, from Goodwyn to Farey to Cauchy, is one of the enduring historical sagas in the theory of numbers. What is known of the time line of this saga is set forth in Table 2.12

TABLE 2.12.    Goodwyn to Farey to Cauchy

| Date | Event |
|------|-------|
| March 5, 1816 | Henry Goodwyn circulates a private printing of his tract "First Centenary of Tables of all Decimal Quotients" |
| April 25, 1816 | Henry Goodwyn, Esq., presents a quarto tract entitled "The First Centenary of a Series of concise and useful Tables of all the complete decimal Quotients which can arise from dividing a Unit or any whole Number less than each Divisor by all Integers from 1 to 1024." to The Royal Society |
| May, 1816 | John Farey publishes a note in The Philosophical Magazine and Journal entitled "On a curious Property of vulgar Fractions" |
| July, 1816 | Farey's note is reproduced in the Bulletin de la Société Philomatique |
| August, 1816 | Cauchy writes "Démonstration d'un Théorème curieux sur les Nombres." in Bulletin de la Société Philomatique proving the property described by Farey |
| 1818 | Henry Goodwyn publishes "The First Centenary of a Series of Concise and Useful Tables of all the Complete Decimal Quotients, which can arise from dividing a unit, or any whole Number less than each Divisor by all Integers from 1 to 1024 To which is now added a Tabular Series of Complete Decimal Quotients for all the Proper Vulgar Fractions of which when in their lowest terms, neither the Numerator nor the Denominator is greater than 100: with the equivalent vulgar fractions prefixed." |
| February 5, 1818 | Henry Goodwin (sic), Esq., donates a quarto tract entitled "The first Centenary of concise and useful Tables of complete Decimal Quotients, &c." to The Royal Society |
| March 6, 1823 | Henry Goodwyn, Esq., presents two octo tracts "A Tabular Series of Decimal Quotients for all proper Vulgar Fractions, in which when in their lowest Terms, neither the Numerator nor the Denominator is greater than 1000." and "A Table of the Circles arising from the Division of a Unit, any other whole Number by all the Integers from 1 to 1024, being all the decimal Quotients that can arise from this source." |
| March 30, 1823 | Henry Goodwyn, Esq., presents "Introduction to a Synoptical Table of English and French Lineal Measures." to the Royal Society |

In his note in The Philosophical Magazine which we will display in full shortly, Farey says he had seen "... some very curious and elaborate Tables of 'Complete decimal Quotients,' calculated by Henry Goodwyn, Esq. of Blackheath, of which he has printed a copious specimen, for private circulation among curious and practical calculators, preparatory to printing of the whole of these useful Tables ...".

Dickson on page 156 of [62] says that the mediant property appears on page five of the table that the 1818 publication added to the 1816 publication. Dickson concludes that the mediant property was also stated in the 1816 tract and that therefore Goodwyn rather than Farey should be mis-credited with the theorem. Dickson notes that Merrifield [179] also credits Goodwyn.

Glaisher [94] observes that the Tabular Series added in the 1818 tract starts with a fresh title-page which is followed by seven pages of introduction, fifteen pages of the table itself and finally fourteen pages of appendix. Glaisher summaries his view of the situation as follows:

The 'First Centenary' (pp. xiv+18) is exactly similar to the first Centenary of 1818; and as the fractions are not arranged in order of magnitude, it contains nothing that in any way suggests either of the properties that form the subject of this paper. It seems pretty clear that no part of the 'Tabular Series' was published previous to 1818; for the title-page to the tract of 1818 runs "The First Centenary ... to which is not added a Tabular Series ...;" and the introduction to the 'Tabular Series'(1818) commences, "Since the 'First Centenary, &c.' and its Introduction were printed, which was in March, 1816, it has appeared to the Calculator ... "

It would thus appear that Mr. Goodwyn published no Table for the conversion of vulgar fractions into decimals, in which the fractions were arranged in order of magnitude, prior to the 'Tabular Series' of 1818; and in this work both the properties are referred to. In the 'Tabular Series' of 1823 only the first is stated. The wording of Mr. Farey's letter implies that he had seen not only the printed specimen of 1816, but also Mr. Goodwyn's manuscript Tables. It is not clear, however, whether Mr.

Farey discovered the property he enunciated without any assistance from Mr. Goodwyn; or whether, Mr. Goodwyn having remarked the property as holding good in the 'Tabular Series,' i.e. when the denominator is 100, Mr. Farey merely extended it to the general case of any denominator. Whoever first began to arrange the fractions in order of magnitude could scarcely fail to notice both properties; and the second, which relates to the difference of two consecutive fractions, would probably present itself first. On the whole, therefore, it seems most probable that only the first extension to the general case was due to Mr. Farey. In none of Mr. Goodwyn's works is any allusion made to Mr. Farey or to Cauchy.

No reference to Charles Haros is made in Glaisher's works. Hardy [124] credits Dickson [62] with the precedence of Haros. The matter could be settled by examining the archive of Goodwyn papers.

## 2.15. The Dispersal of Goodwyn's Archive

Glaisher [94] reports that the following letter was found in one of the 1818 copies of 'First Centenary' in the Cambridge University Library:

> September 16$^{th}$, 1831
>     Mrs. Catherine Goodwyn presents to the Library of the University of Cambridge a complete set of the works of her late father, Henry Goodwyn, Esq., of Blackheath, Kent. Royal Hill, Greenwich.

In the entry under Tables entry in Volume VII of the Fourth Division of "The English Cyclopedia" published in 1868 and attributed by Glaisher to De Morgan, the following is found:

> Mr. Goodwyn (of Blackheath) was an indefatigable calculator; and the preceding Tables are the only ones of the kind published. His manuscripts, an enormous mass of similar calculations, came into the possession of Dr. Olinthus Gregory,

and were purchased by the Royal Society at the sale of his books in 1842.

And yet in Volume LXV of The Philosophical Magazine and Journal of 1825 there is a note in the Intelligence and Miscellaneous Articles Section signed by Olinthus Gregory and dated November, 1824. The note is titled "The British Museum - Mr. Goodwyn's Manuscripts" and starts out:

> Those who are interested in mathematical computations, and the tabulation of their results, for practical purposes, will learn with pleasure that the curious and extensive Tables of the late Henry Goodwyn, Esq. of Blackheath, have, by the advice of Dr. Gregory, Professor of Mathematics in the Royal Military Academy, been deposited by Mr. Goodwyn's family in the library of the British Museum.

The note concludes:

> Mr. Goodwyn's family, anxious to consign these manuscripts of their revered relative to some institution where they may be occasionally consulted by the friends and promoters of mathematical science, do now, with the consent of the trustees of the British Museum, deposit them in the library of that magnificent national institution.

Glaisher says in [94] that the Royal Society knows nothing of the Goodwyn manuscripts, but that two copies each of the 1818 and 1823 publications are in The Cambridge University Library.

Glaisher was a prolific writer and table analyzer. The following is from a 1947 review by Leslie Comrie of John Thomson's Table of Twelve-Figure Logarithms:

> It may not be without interest to record that when the present writer tried in 1926 to see these tables at the library of the Royal Astronomical Society, he was informed that they were out on loan. After persistent efforts by the Secretary, they were

received two years later from Dr. Glaisher, who had them for 47 years! But for this intervention, they might have gone to a bookseller when his library was disposed of after his death in 1928. Now after three quarters of a century, they have once more been useful as an independent check on a published table. [**46**]

In a footnote in [**93**] Glaisher writes

> Since this paper was communicated to the Society I have written and sent to the *Philosophical Magazine* a detailed historical account of the two theorems, with demonstrations of them. J.W.L.G., February 20, 1879.

He is very likely referring to [**94**] which has an 8-page history of the theorems but no mention of the work of Charles Haros.

The following Query appeared in the April, 1943, issue of *Mathematical Tables and Other Aids to Computation*:

> **2.** SCARCE MATHEMATICAL TABLES. – In what libraries of the world, public or private, may the following books be found:
> ...
>
> E. [Henry Goodwyn], *A Table of Circles arising from the Division of a Unit or any other Whole Number, by all the Integers from 1 to 1024, being all the pure Decimal Quotients that can arise from this source.* London, 1823 v, 118 p. Published Anonymously.

Tables listed as *A.* through *D.* were not by Goodwyn.

In the Recent Mathematical Tables of the previous, inaugural issue of MTAC there had been a paragraph describing the location of copies of four of Goodwyn's tables. The paragraph starts out with the sentence "These publications are excessively rare."

TABLE 2.13.   Conjectured Locations of Goodwyn's Manuscripts

| Ref. | Pub. | Location |
|------|------|----------|
| [105] | 1816 | Yale University Library, New Haven |
|       |      | Goldsmith's Library, University of London, London |
|       |      | British Library, London |
|       |      | Royal Society, London (1825 Catalogue) |
| [106] | 1818 | Brown University, Providence |
|       |      | Royal Society, London (1825 Catalogue) |
|       |      | Royal Society, London (1839 Catalogue, two copies) |
| [107] | 1820(?) | Royal Society, London |
| [109] | 1821 | Royal Society, London (1839 Catalogue) |
| [111] | 1823 | Library of the Faculty of Advocates, Edinburgh |
|       |      | Crawford Library of the Royal Observatory, Edinburgh |
|       |      | Library of the University of Edinburgh, Edinburgh |
|       |      | Library of the University of Cambridge, Cambridge |
|       |      | Brown University, Providence |
|       |      | Harvard University, Cambridge |
|       |      | Royal Society, London (1839 Catalogue) |
|       |      | Library of L.J. Comrie |
| [112] | 1823 | John Crerar Library, Chicago |
|       |      | Royal Society, London (1839 Catalogue) |
|       |      | Cambridge University Library, Cambridge (2 copies) |

Table 2.13 summarizes the current locations of original copies of Goodwyn's tables gleaned from these sources and and research at the Royal Society. The date of the table of logarithms is based on a letter found in the manuscript and is uncertain.

As an indicator of the tectonic shifts taking place in the historiography of mathematics, in addition being available in downloadable digital form on Google Books, a printed reproduction of Harvard University's copy of Goodwyn's 1823 manuscript is available on-line for $19.31.

## 2.16. Goodwyn's Publications

Table 2.14 lists some of Goodwyn's publications. Much like John Farey and other gentlemen scientists of the day, he didn't confine himself to a narrow speciality. That said, it is easy to see Goodwyn's experience as a brewer in his papers.

## 2.17. "On the Quotient arising from the Division of an Unit..."

*On the Quotient arising from the Division of an Unit by prime Numbers. By*

*H. Goodwyn, Esq.*

To Mr. NICHOLSON.

SIR,

THE following account of the quotients arising from the division of an unit, &c. by prime numbers, being, I believe, perfectly new, and promising' to be very useful, is very much at your service; and if you think it worthy a place in your Journal, it may induce, the publication of a small table prepared for a farther elucidation of the subject.

I am, SIR,

Respectfully your's,

*East Smithfield, Oct.* 1800.                    H. GOODWYN.

The quotient of an unit, divided by the prime number 17, will consist of 16 places of ligures, forming a complete circulating decimal. If the numbers, 2, 3, 4, &c to 16, be divided by the same prime number, each respective quotient will still consist of 16 places of circulating decimals. Thus far the property of like divisions. has been ascertained by various writers on decimal arithmetic, &c.

TABLE 2.14. Some of Goodwyn's Publications

| Year | Title |
|------|-------|
| 1796 | The Brewers Assistant, containing a variety of tables, calculated to find ...the value, quantity, weight, &c. of the principal articles purchased, expended, sold, or retained in a brewing trade [98] |
| 1800 | On the new Measures of France [99] |
| 1800 | A Section and Description of a Machine that will raise a Body of Water to any Height, not exceeding the Height of a Column that will counterbalance the Pressure of the Atmosphere (say 30 Feet) by the Descent of Part of the same Body of Water, through a somewhat greater height, and aided by the Pressure of the Atmosphere [100] |
| 1801 | On raising Water by the Engine [102] |
| 1801 | Curious properties of prime numbers, taken as the divisors of unity [103] |
| 1801 | Construction and Use of an universal Table of Interest[101] |
| 1803 | A Table to compare a new System of English with the new System of French Measures and Weights ... [104] |
| 1816 | The first centenary of a series of ...tables of all the complete Decimal Quotients which can arise from dividing a unit, or any whole number less than each divisor by all integers from 1 to 1024. [Preceded by an introduction, etc.] [105] |
| 1818 | The first centenary of a series of ...tables of all the complete Decimal Quotients which can arise from dividing a unit ... by all integers from 1 to 1024. To which is now added a tabular series of complete decimal quotients for all the proper vulgar fractions, etc. [106] |
| 1820 | A very concise, yet strictly accurate method, for finding the interest of any given sum : at any given rate, for any given number of days [108] |
| 1821 | Introduction to a synoptical table of English and French lineal measures [109] |
| 1823 | A Table of the Circles arising from the Division of a Unit, any other whole Number by all the Integers from 1 to 1024, being all the decimal Quotients that can arise from this source [111] |
| 1823 | A Tabular Series of Decimal Quotients for all proper Vulgar Fractions, in which when in their lowest Terms, neither the Numerator nor the Denominator is greater than 1000 [112] |
| 1823 | A diagonal table to facilitate the comparison of measures of capacity, which result naturally from the proposed imperial gallon [110] |
| 1825 | A Table of Circles, from which knowing the diameters, the areas, circumference, and sides of equal squares, are found. [117] |

But at least one very curious, concise, and useful property attached to similar divisions in general, yet remains to be unfolded. It is this – that the quotient arising from the first division, virtually represents the quotient of every other division above-mentioned.

And in like manner the quotient arising from the division of an unit by every other whole number, less than the divisor, will commence with a different figure in the first quotient, and will circulate to that figure again.

And thus the complete quotient arising from the division of each whole number, less than the divisor, in that division may be expressed on the first quotient, by placing the respective dividends over their first quotient figure. Thus

| 1 | 10 | 15 | 14 | 4 | 6 | 9 | 5 | 16 | 7 | 2 | 3 | 13 | 11 | 8 | 12 | dividends |
|---|----|----|----|---|---|---|---|----|---|---|---|----|----|---|----|-----------|
| 0 | 5  | 8  | 8  | 2 | 3 | 5 | 2 | 9  | 4 | 1 | 1 | 7  | 6  | 4 | 7  | quotients |

This disposition of the dividend and first quotient enables us to find, by inspection, the complete decimal quotient. or expression for a vulgar fra8ion, whole numerator or dividend is any given whole number, between 1 and 17, and whole denominator or divisor is 17. The quotient of $\frac{2}{17}$, $\frac{3}{17}$, and $\frac{4}{17}$ seen in the first elucidating arrangement above, and perfectly coincides with this last. By this, if I want the complete decimal quotient of $\frac{16}{17}$, I have only to search for the number 16 in the line of dividends, and under it is the first figure of the circulating decimal that will comprize (sic) complete quotient of $\frac{16}{17}$ viz .9411764705882352, and the same of the other dividends.

But what authorises the above property to be termed curious, concise, and useful, is, that it does not attach to the prime number 17 only, but under certain laws ia equally, applicable to all prime and multiples of prime numbers whatever.

## 2.18. Goodwyn and the Mediant Property

An enduring question is whether or not Goodwyn knew about the mediant property of his sequence of vulgar fractions before Farey pointed it out and, if he did, whether he discovered it himself or if, perhaps, he knew, directly or indirectly, about Haros' mediant when he published his first table of vulgar fractions in 1816.

Goodwyn was a sufficiently talented table maker that he wouldn't have set out to compute a table with 304,193 entries without some map of the road ahead. Research in the Goodwyn archives at the Royal Society, Collection MS/781, reveals that Goodwyn, as we suspect Flitcon before him, developed on his own Euler's product formula for computing the number of entries in his table. This tells him how big his table will be.

Figure 2.11 is a page of Goodwyn's 1816 table. All the fractions in the Farey series of order 100 are in the table but not in magnitude order. Based on work sheets in Goodwyn's Royal Society archive he computed the vulgar fractions for each denominator by sieving all the positive integers less than the denominator.

Goodwyn's 1818 publication was the 1816 publication "To which is now added a tabular series of complete decimal quotients for all the proper vulgar fractions of which, when in their lowest terms, neither the numerator, nor the denominator, is greater than 100," namely his recomputation of Haros' table.

Based on Glaisher's quotations from the 1818 publication [94], Goodwyn was fully aware of the mediant at this time and no doubt used it to build his 1823 table. Glaisher goes on to say

> The wording of Mr. Farey's letter implies that he had seen not only the printed specimen of 1816, but also Mr. Goodwyn's manuscript Tables. It is not clear, however, whether Mr. Farey discovered the property he enunciated without any assistance from Mr. Goodwyn; or whether, Mr. Goodwyn having remarked the property as holding good in the 'Tabular Series,' i.e. when the denominator is 100, Mr. Farey merely extended it to the general case of any denominator. ...In none of Mr. Goodwyn's works is any allusion made to Mr. Farey or to Cauchy.

**29**
P

·965517241379931
034482755862068

| | | |
|---|---|---|
| 28 | ·965517 | 1 |
| 27 | ·931034 | 2 |
| 26 | ·896551 | 3 |
| 25 | ·862068 | 4 |
| 24 | ·827586 | 5 |
| 23 | ·793103 | 6 |
| 22 | ·758620 | 7 |
| 21 | ·724137 | 8 |
| 20 | ·689655 | 9 |
| 19 | ·655172 | 10 |
| 18 | ·620689 | 11 |
| 17 | ·586206 | 12 |
| 16 | ·551724 | 13 |
| 15 | ·517241 | 14 |

**30**
$2 \cdot 3 \cdot 5$

| | | |
|---|---|---|
| 29 | ·96 | 1 |
| 23 | ·76 | 7 |
| 19 | ·63 | 11 |
| 17 | ·56 | 13 |

**31**
P

·967741935483870
032258064516129

| | | |
|---|---|---|
| 30 | ·967741 | 1 |
| 29 | ·935483 | 2 |
| 28 | ·903225 | 3 |
| 27 | ·870967 | 4 |
| 26 | ·838709 | 5 |
| 25 | ·806451 | 6 |
| 24 | ·774193 | 7 |
| 23 | ·741935 | 8 |
| 22 | ·709677 | 9 |
| 21 | ·677419 | 10 |
| 20 | ·645161 | 11 |
| 19 | ·612903 | 12 |
| 18 | ·580645 | 13 |
| 17 | ·548387 | 14 |
| 16 | ·516129 | 15 |

**32**
$2^5$

| | | |
|---|---|---|
| 31 | ·96875. | 1 |
| 29 | ·90625. | 3 |
| 27 | ·84375. | 5 |
| 25 | ·78125. | 7 |
| 23 | ·71875. | 9 |
| 21 | ·65625. | 11 |
| 19 | ·59375. | 13 |
| 17 | ·53125. | 15 |

**33**
$3 \cdot 11$

| | | |
|---|---|---|
| 32 | ·96 | 1 |
| 31 | ·93 | 2 |
| 29 | ·87 | 4 |
| 28 | ·84 | 5 |
| 26 | ·78 | 7 |
| 25 | ·75 | 8 |
| 23 | ·69 | 10 |
| 20 | ·60 | 13 |
| 19 | ·57 | 14 |
| 17 | ·51 | 16 |

**34**
$2 \cdot 17$

·70588235
29411764

| | | |
|---|---|---|
| 33 | ·970588 | 1 |
| 31 | ·911764 | 3 |
| 29 | ·852941 | 5 |
| 27 | ·794117 | 7 |
| 25 | ·735294 | 9 |
| 23 | ·676470 | 11 |
| 21 | ·617647 | 13 |
| 19 | ·558823 | 15 |

**35**
$5 \cdot 7$

| | | |
|---|---|---|
| 34 | ·9714285 | 1 |
| 33 | ·9428571 | 2 |
| 32 | ·9142857 | 3 |
| 31 | ·8857142 | 4 |
| 29 | ·8285714 | 6 |
| 27 | ·7714285 | 8 |
| 26 | ·7428571 | 9 |
| 24 | ·6857142 | 11 |
| 23 | ·6571428 | 12 |
| 22 | ·6285714 | 13 |
| 19 | ·5428571 | 16 |
| 18 | ·5142857 | 17 |

**36**
$2^2 \cdot 3^2$

| | | |
|---|---|---|
| 35 | ·972 | 1 |
| 31 | ·861 | 5 |
| 29 | ·805 | 7 |
| 25 | ·694 | 11 |
| 23 | ·638 | 13 |
| 19 | ·527 | 17 |

**37**
P

| | | |
|---|---|---|
| 36 | ·972 | 1 |
| 35 | ·945 | 2 |
| 34 | ·918 | 3 |
| 33 | ·891 | 4 |
| 32 | ·864 | 5 |
| 31 | ·837 | 6 |
| 30 | ·810 | 7 |
| 29 | ·783 | 8 |

FIGURE 2.11.  Tables in Goodwyn's 1816 Publication

Let us consider first the possibility that Goodwyn knew of Haros' work and used Haros' mediant.

*Scenario #1: Goodwyn Uses Haros' Mediant*

On February 24, 1803, Goodwyn presented a folio to the Royal Society entitled "A Table to compare a new System of English with the new System of French Measures and Weights" [**104**]. While he been in the habit of publishing every three or four years, Goodwyn went silent for 13 years, until 1816, when he published the first part of "The First Centenary of a Series of Concise and Useful Tables of all the Complete Decimal Quotients, which can arise from dividing a unit, or any whole Number less than each Divisor by all Integers from 1 to 1024" [**105**]. This was the tract in which John Farey noticed a curious property. Starting in 1816 he once again published every three or four years.

In 1813 a multi-volume work appeared with the imposing title, *Pantologia. A New Cyclopedia Comprehending a Complete Series of Essays, Treatises, and Systems, Alphabetically Arranged; with a General Dictionary of Arts, Sciences, and Words: the Whole Presenting a Distinct Survey of Human Genius, Learning, and Industry, Illustrated with Elegant Engravings; those of Natural History being from Original Drawings by Edwards and Others, and Beautifully Coloured after Nature.* appeared. The editors of this work were John Mason, Olinthus Gregory and Newton Bosworth "assisted by other gentlemen of eminence in different department of literature." Gregory wrote the entry under ARITHMETIC in which we find the following:

> As another good approximating rule, we shall add that discovered by Haros; and which is,
> Find the nearest power to the given number, and multiply the difference between it and the given number by double its root for a dividend. – And for a divisor, multiply the nearest power by double the index of its root, and add to or take from the product, as the power is too little or too great, the above difference multiplied by one less than the index. Then the root of the nearest power increased or diminished, as the case may require, by the quotient obtained by the division, will be the true root nearly.
> N. B. Two operations are generally necessary, in order to have the answer true to 6 or 7 figures; and the result of the first must always be used as the root in the second.

According to Hutton [**138**], Haros was admired in Britain. One of the individuals familiar with Haros' work was Olinthus Gregory. As we saw above, Olinthus Gregory was a close friend of Henry Goodwyn.

Is it possible that in researching Haros' work on approximation, Gregory could have come upon a copy of Haros' 1802 paper? And if Gregory was in possession of Haros' paper might he have provided it to Goodwyn as a source document for Goodwyn's 1803 paper comparing the English system of weights and measures to the new French system of weights and measures? We know that Goodwyn was an avid table maker. With Haros' mediant algorithm in hand might he not imagine he could reach for 1024?

The table that Goodwyn starts but never finishes was successfully completed 150 years later. The project would be supervised by a Cambridge second wrangler in mathematics and a team of eminently qualified human computers. Goodwyn was a retired brewer working alone. Are we to believe that he thought he could make a list of all 318,965 of the fractions he proposed to study without an algorithm of some sort?

One must also observe that while Haros' mediant algorithm is not mentioned in the 1816 version of Goodwyn's partial table, it is nonetheless described in detail two years later in the 1818 update, after Farey's curious property letter was published in *The Philosophical Magazine and Journal*.

Is it possible that Farey didn't observe the curious property but learned about the property from Goodwyn, Gregory or somebody else? Or is it possible that Farey blew Goodwyn's cover by really noticing the mediant property in Goodwyn's table and naively reporting his observation in a letter to the *The Philosophical Magazine and Journal*? Finally, could Gregory have purged Goodwyn's papers of any content related to Haros while these papers were in his possession?

*Scenario #2: Goodwyn Discovers the Mediant*

Now let's consider the possibility that Goodwyn came up with his own algorithm for generating the 304,193 entries in his table. The above mentioned research in the Royal Society's Goodwyn archive turned up the document pictured in Figure 2.12.

Take any three consecutive Vulgar Fractions in the Tabular Series of (Complete Decimal Quotients). The Sum of the Denominators of the $1^{st}$ and $3^{rd}$ will always be divisible by the Denominator of the $2^{nd}$ without Remainder. Also the Sum of the Numerator of the $1^{st}$ and $3^{rd}$ will also (sic) be divisible by the Numerator of the $2^{nd}$ without Remainder.

If the $2^{nd}$ or middle Denominator be multiplied by the quotient arising from the Sum of the $1^{st}$ and $3^{rd}$ Numerator divided by the second, the Product will equal the Sum of the $1^{st}$ and $3^{rd}$ Denominator.

The second paragraph in symbols says

$$d_2 \frac{n_1 + n_3}{n_2} = d_1 + d_3,$$

that is

$$\frac{n_2}{d_2} = \frac{n_1 + n_3}{d_1 + d_3},$$

Figure 2.13 is a close-up of the postal time stamp on Figure 2.12. If one runs through the ten possibilities for the final digit one can with some reasonableness eliminate each one except two, three and seven. If the digit is a two or a three, then Goodwyn knew about the mediant property in 1812, four years before Farey discovered it in 1816 in a table Goodwyn had not yet published.

Furthermore, in contrast to Glaisher's characterization of Goodwyn's mediant, this is a perfectly general statement of the mediant property that is independent of any particular denominator maximum.

## 2.19. Decimalization of the Pound Sterling

The French franc was decimalized in 1795 and as we've seen the Farey sequence first appeared in a 1802 table that Charles Haros constructed to help people translate between the old fractional system and the new decimal system. Due to France's conversion to a decimal system, decimalization was topic of heated discussion in Henry Goodwyn's Britain. Since the British pound sterling was not decimalized until 1971 the anti-decimalization forces clearly held sway.

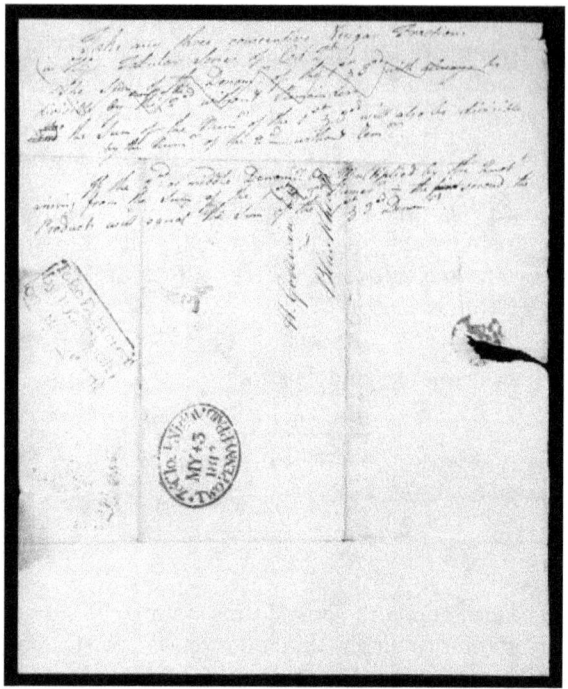

FIGURE 2.12.   Goodwyn's Statement of the Mediant

This may account for why Goodwyn includes the following note at the end of the introduction to his 1816 publication:

> Lest it should be inferred from the appearance of this Specimen at the time when a Bill is under the consideration of Parliament for the *Equalisation of the* WEIGHTS *and* MEASURES *of the Kingdom*, that the Calculator is an advocate for a Decimal Division, he takes the liberty most respectfully to offer his opinion – that, for common use, the Standards of both should be derived from 2 and its *powers*; and, where any intervening Weights or Measures are necessary, that they should be expressed by *products* of that number. ... This division of

FIGURE 2.13. Time Stamp on Goodwyn's Statement of the Mediant

the Weights and Measures, the Calculator is happy to find,
is nearly the same as the one which has been recommended
to the Committee of the House of Commons, by gentlemen of
such acknowledged abilities as Mr. Professor Playfair, and the
present Secretary to the Royal Society, Dr. Wollaston.

On a quiet evening one might be given to wondering why anyone would
possibly think that a modest table of decimal quotients was an argument in
favor of decimalization of the pound sterling; that is unless there was a similar
table in circulation that was explicitly constructed to aid a decimalization effort.

## 2.20. John Farey, Geologist and Musicologist

John Farey was an extraordinary geologist before geology existed [80]. He worked closely with William Smith who is credited with founding the discipline of geology. Farey was also a polymath who dabbled in many scientific fields; one of these was mathematics.

One Farey's interests beside geology and mathematics was music and in particular the intersection of music and mathematics. Perhaps this is what gave him a keen eye for patterns and relationships between successive notations. At any rate, the story goes that he discerned the mediant property in Goodwyn's tables and published the following very short note about his observation in *The Philosophical Magazine and Journal* [71].

## 2.21. "On a Curious Property of Vulgar Fractions"

LXXIX. *On a curious Property of vulgar Fractions. By*

*Mr.* J. Farey, *Sen.*
*To Mr. Tilloch.*

Sir, — On examining lately, some very curious and elaborate Tables of "Complete decimal Quotients," calculated by Henry Goodwyn, Esq. of Blackheath, of which he has printed a copious specimen, for private circulation among curious and practical calculators, preparatory to the printing of the whole of these useful Tables, if sufficient encouragement, either public or individual, should appear to warrant such a step : I was fortunate while so doing, to deduce from them the following general property; viz.

If all the possible vulgar fractions of different values, whole greatest denominator (when in their lowest terms) does not exceed any given number, be arranged in the order of their values, or quotients; then if both the numerator and the denominator of any fraction therein, be added to the numerator and the denominator, respectively, of the fraction next but one to it (on either side), the sums will give the fraction next to it; although, perhaps, not in its lowest terms.

For example, if 5 be the greatest denominator given; then are all the possible fractions, when arranged, $\frac{1}{5}, \frac{1}{4}, \frac{1}{3}, \frac{2}{5}, \frac{1}{2}, \frac{3}{5}, \frac{2}{3}, \frac{3}{4}$ and $\frac{4}{5}$; taking $\frac{1}{3}$ as the given fraction, we have $\frac{1 \pm 1}{5 + 3} = \frac{2}{8} = \frac{1}{4}$ the next smaller fraction then $\frac{1}{3}$; or, $\frac{1 \pm 1}{3 + 2} = \frac{2}{5}$, the next larger fraction to $\frac{1}{3}$. Again, if 99 be the largest denominator, then, in a part of the arranged Table, we should have $\frac{15}{52}, \frac{28}{97}, \frac{13}{45}, \frac{24}{83}, \frac{11}{38}$, &c.; and if the third of these fractions be given, we have $\frac{15 \pm 13}{52 + 45} = \frac{28}{97}$ the second: or $\frac{13 \pm 11}{45 + 38} = \frac{24}{83}$ the fourth of them : and so in all the other cases.

I am not acquainted, whether this curious property of vulgar fractions has been before pointed out?; or whether it may admit of any easy or general demonstration ?; which are points on which I should be glad to learn the sentiments of some of your mathematical readers; and am

Sir,

Your obedient humble servant,

J. FAREY,

Howland-street.

Farey refers to the table in which he noticed the curious property as "a copious specimen, for private circulation among curious and practical calculators, preparatory to the printing."

A search of the Goodwyn archive at the Royal Society turned up a "List of Personages to whom Copies of Specimen of Complete decimal Quotients have been presented." The title of this list is shown in Figure 2.14. The list spans three pages, the last page of which is shown in Figure 2.15. On this last page we see Farey's name mentioned three times: once as "Farey Jn Sen Mineral Surveyor," next as "Farey J," and finally in association with the entry for Benjamin Bevan as "via Farey Sen."

We can I think conclude that Farey and Goodwyn knew one another and that Farey came by the table he examined directly from Goodwyn.

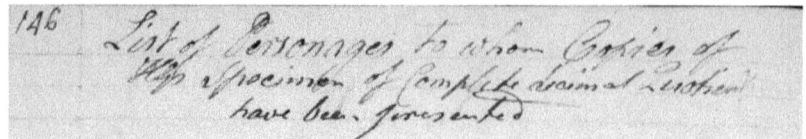

FIGURE 2.14.   Title on List of Recipients of Specimen

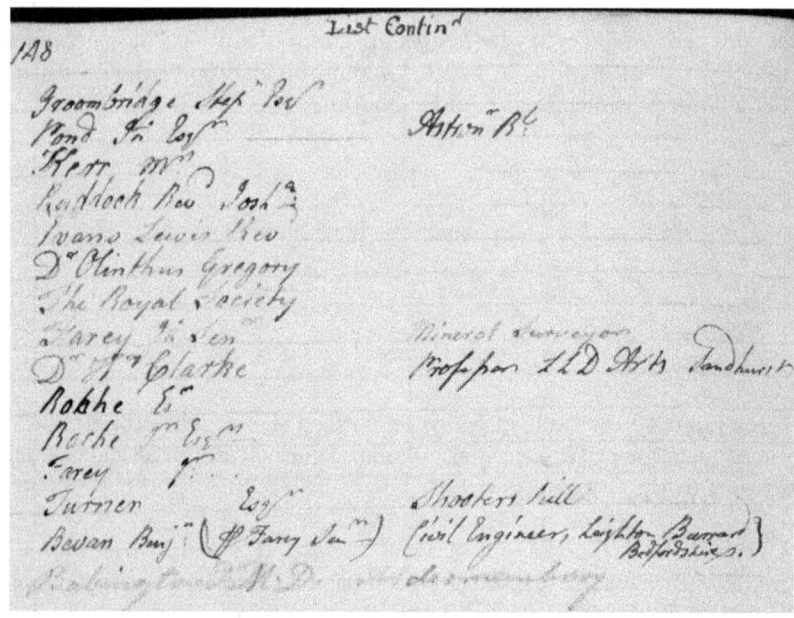

FIGURE 2.15.   Farey's Name on List of Recipients of Specimen

Figure 2.16 is a clip from Smith's New Map of London published in 1860. Howland Street where Farey lived is in the upper left and East Smithfield where Goodwyn was in 1801 in the lower right. According to Google maps it's a seventy-minute, $3\frac{2}{5}$ mile walk between the two; roughly Regency Park to the Tower of London.

FIGURE 2.16.   Howland Street and East Smithfield in 1860

Farey's letter appeared in the May 1816 issue Alexander Tilloch's *Philosophical Magazine and Journal*. Two months later the following synopsis of Farey's letter appeared in the *Bulletin des Sciences, par la Société Philomatique de Paris*:

> *A curious property of ordinary fractions.*
>
> If one arranges in order of size all possible fractions whose largest denominator (when one has reduced them to their simplest expression) does not exceed a given number, and next, one adds the numerator and the denominator of one of these fractions respectively to the numerator and the denominator of the fraction that precedes it or follows it by two places, one will obtain the preceding fraction or the one that follows next, which, however, may not be reduced to its simplest expression.
>
> Example: Let 7 be the largest given denominator. Here are all the possible fractions arranged in order of magnitude:
>
> $$\frac{1}{7}, \frac{1}{6}, \frac{1}{5}, \frac{1}{4}, \frac{2}{7}, \frac{2}{3}, \frac{3}{7}, \frac{1}{2}, \frac{4}{7}, \frac{3}{5}, \frac{2}{3}, \frac{5}{7}, \frac{3}{4}, \frac{4}{5}, \frac{5}{6}, \frac{6}{7}, \frac{1}{1}$$
>
> Let us take $\frac{1}{6}$, and we will have $\frac{1+2}{7+5} = \frac{2}{12} = \frac{1}{6}$ a fraction that is immediately smaller than $\frac{1}{5}$ and next $\frac{1+2}{5+7} = \frac{3}{12} = \frac{1}{4}$ a fraction that is immediately larger than $\frac{1}{5}$. If one takes $\frac{4}{5}$ one

will have $\frac{4+5}{5+7} = \frac{1}{4}$ and $\frac{4+6}{5+7} = \frac{5}{6}$ for the fraction that is immediately smaller or immediately larger than $\frac{4}{5}$. The English writer states that he is unaware if anyone reported this previously.

The article two before this note *Bulletin des Sciences, par la Société Philomatique de Paris* is by Poisson. Poisson's article is dated June 1816. Based on this, Cauchy's paper can be dated to July 1816. In his paper, Cauchy starts out referring to the note on Farey's observation as being in the previous edition of the *Bulletin* so we can conclude that the note appeared in June 1816, two months after it appeared in Tilloch's journal. The note itself does not name Farey but refers only to an "English writer."

This excerpt was taken from the volume containing all monthly editions of the *Bulletin* for 1816 and 1817, not from the individual *Bulletin* in which it appeared so it is possible that what appeared in the monthly bulletin was different from what ended up in the volume, but given the small size and minor importance of the note this is I think unlikely.

With respect to the fact that Farey's name is not included in the note and yet is used by Cauchy, Glaisher says the following:

> It is clear, from the first paragraph of (Cauchy's) paper, that he must have referred to Mr. Farey's original letter in the Philosophical Magazine, since, as has been mentioned, Mr. Farey's name does not occur in the account in the *Bulletin*. [**94**]

### 2.22. "Proof of a Curious Theorem Regarding Numbers"

Here's the paper by Cauchy [**34**] where the name of John Farey became bound to a sequence of vulgar fractions.

PROOF
OF A CURIOUS THEOREM REGARDING NUMBERS.

In an issue of the *Bulletin de la Société Philomathique*, there is an enunciation of a remarkable property of ordinary fractions observed by M. J. Farey: This property represents nothing more than a simple corollary of a curious theorem that I will begin by laying out.

THEOREM. – If, having arranged all irreducible fractions whose denominator does not exceed a given whole number in order of their size, one then selects two consecutive fractions at will from the series thus formed, their denominators will be mutually coprime, and their difference will be a new fraction whose numerator will be one.

PROOF. Let $\frac{a}{b}$ be the smaller of the two fractions under consideration, and $n$ the whole number that is given. In addition, let $a'$ and $b'$ be the largest whole numbers that can be assigned to the variables $x$ and $y$ in the indeterminate equation,

$$bx - ay = 1,$$

always assuming that $b' < n$. Since the fraction $\frac{a}{b}$ is irreducible by definition, and the value of $b'$ conforming to the equation

$$ba' - ab' = 1,$$

$b$ and $b'$ will necessarily be mutually coprime, and, in addition, one will have

$$\frac{a'}{b'} - \frac{a}{b} = \frac{1}{bb'}.$$

The fraction, $\frac{a'}{b'}$ will thus show the properties enunciated in the theorem with respect to the fraction $\frac{a}{b}$; and, in order to prove this same theorem, it will suffice to prove that, among all irreducible fractions whose numerators to not exceed $n$, the one that is next greater than $\frac{a}{b}$ is precisely $\frac{a'}{b'}$ . One arrives at demonstrating this in the following fashion.

The different values of $y$ given by equation (1) form the arithmetic series

$$\ldots, b' - 2b, b' - b, b', b' + b, b' + 2b, \ldots;$$

and, since $b'$ is the largest of these values that is included in $n$, one necessarily has

$$n < b' + b.$$

Now, let $\frac{f}{g}$ be an irreducible fraction larger than $\frac{a}{b}$ taken from among those whose denominators do not exceed $n$. If, for the sake of abbreviation, one states,

$$bf - ag = m$$

then one will have

$$\frac{f}{g} - \frac{a}{b} = \frac{m}{bg}.$$

Thus, the difference between the fractions $\frac{f}{g}$ and $\frac{a}{b}$ will be generally expressed by $\frac{m}{bg}$; and, if one assigns a constant value to $m$ while allowing $g$ to vary, this difference will have the minimum possible value when $g$ has the maximum possible value. Additionally, the different values for $g$ satisfying equation (2) are clearly comprised by the arithmetic series

$$\ldots, mb' - 2b, mb' - b, mb', mb' + b, mb' + 2b, \ldots,$$

whose term $mb' + b$ is equal to or greater than $b' + b$ and is thus greater than $n$; and, since $g$ may not exceed $n$, it is clear that it will additionally be equal to the term $mb'$; from whence it follows that the fraction $\frac{m}{bg}$ cannot become less than

$$\frac{m}{mb'b} = \frac{1}{b'b}.$$

Thus, among all the fractions that are greater than $\frac{a}{b}$, and whose denominator does not exceed $n$, the smallest is the one whose difference from $\frac{1}{bb'}$ is equal to $\frac{1}{bb'}$ , that is to say, the fraction $\frac{a'}{b'}$.

COROLLARY. – If one selects any three consecutive fractions from among those under consideration in the theorem, designating these three fractions as

$$\frac{a}{b}, \frac{a'}{b'}, \frac{a''}{b''},$$

then one will have

$$a'b - ab' = 1, \quad a''b' = a'b'' = 1,$$

and as a result

$$\frac{a + a''}{b + b''} = \frac{a'}{b'}.$$

from which one then concludes that

$$\frac{a + a''}{b + b''} = \frac{a'}{b'}.$$

This last equation is none other than the analytic expression for the property observed by M. J. Farey.

By the time he wrote *Cours D'Analyse de l'École Royale Polytechnique* [**35**], Cauchy had generalized this theorem slightly:

THÉORÈME. *Soient* $b, b', b'', \ldots$ *plusiers quantités de même signe* $n$, *et* $a, a', a'', \ldots$ *des quantités quelconques en nonbre égal á celui des premiéres. La fraction*

$$\frac{a + a' + a'' + \&c\ldots}{b + b' + b'' + \&c\ldots}$$

*sera moyene entre les suivantes*

$$\frac{a}{b}, \frac{a'}{b'}, \frac{a''}{b''}, \&c. \ldots$$

## 2.23. Delambre and Tilloch Weigh In

On page 345 of the 1817 issue of Alexander Tilloch's *Philosophical Magazine and Journal*, Tilloch reports receiving an 80 quarto page report from M. Le Chevalier Delambre, the Perpetual Secretary of the Mathematical Department of the Institute of France reporting on recent activities of the department. Both Farey's letter and Cauchy's proof seem to have been referenced in Delambre's report but the report itself was too big to be reprinted *in toto* in Tilloch's journal so Tilloch summarizes:

> The article of Mr. Farey's to which we have just alluded, and to which the ingenious secretary (Delambre) refers with no more than the respect which it merits, appeared originally in the Philosophical Magazine, vol. xlvii. p. 385, and refers to a curious property of vulgar fractions, which Mr. Farey was first to discover and communicate to the world.

I have not located a copy of Delambre's report but it is unlikely that if Delambre had openly scoffed at Farey's paper Tilloch would have mentioned Delambre's reference to it. A second possibility is that Delambre's denigration

of the paper may have been so French, whether in English or French, that
Tilloch was $180^o$ out of phase. But a final possibility is that Delambre, like
Cauchy, didn't connect Farey's paper with Haros' work 15 years earlier.

Farey makes no inventive claim on the series of vulgar fractions that now
bear his name nor does he suggest that he proved any properties of the series. It's the mathematical heavies – Cauchy, Delambre and Tilloch – that are
pushing the series onto Farey's resume.

By the way, the letter by S.A. to which Farey refers is dated August 31,
1816, and is entitled "On Vulgar Fractions."

> These rules say that if any four numbers be in arithmetical
> proportion, the sum of the means will always be equal the sum
> of the extremes. For example, since
> $$\frac{15}{52}, \frac{28}{97}, \frac{13}{45}, = \frac{30}{104}, \frac{28}{97}, \frac{26}{90},$$
> both the numerators and the denominators are in arithmetic
> progression so by the rules of arithmetical proportion
> $$\frac{130+26}{104+90} = \frac{28\,2}{97\,2} = \frac{56}{194}.$$

The very next month Tilloch added the following footnote to a report from
Delambre on the 1816 goings on in the mathematics section of the Institute of
France [**59**].

> I am happy in being able at length to announce the speedy pub-
> lication of the first Part of the very useful Set of Arithmetical
> Tables, by Henry Goodwyn, esq, which were first introduced to
> the notice of my readers, in a communication from his friend:
> Mr. Farey sen.*, inserted p. 385 of the xlviith volume. Ev-
> ery one used to calculations must have experienced the labour
> attending the reduction of vulgar fractions to their equivalent
> decimal fractions, although the denominators may be small, as
> under 100, if *many* places of figures are wanted; and the danger
> there is of making mistakes in such cases. And few persons are
> much versed in figures, without having very frequently seen

the still greater labour and difficulties attending the reverse of the above process; viz. the reducing of of decimal fraction to its equivalent vulgar fraction, when such be correctly practicable, or otherwise, of finding *the nearest* vulgar fraction thereto expressed in small numbers;as for instance, less than 100; in either of its terms. Both these operations, Mr. Goodwyn's Tables are calculated to perform by mere inspection; and in the latter particular, especially, to save much time to calculators, as well as secure greater accuracy to their results.

But Mr. Goodwyn goes a great deal further, as the title of his intended work imports; viz. a "First Centenary of a Series of complete Decimal Quotients;" because, this Table will admit, the taking out by inspection of the *complete value* in each case -of every fraction between $\frac{1}{99}$ and $\frac{98}{99}$;that is, with as many places of decimals, as is requisite, either for the same terminating or circulating, as the case may : showing all the various *decimal circles* belonging to each division under 1025.

The elaborate Table of the same kind, intended to follow, if the present intended ones are favourably received, will carry Mr. Goodwyn's series of complete decimal quotients to $\frac{998}{999}$ or even to $\frac{1023}{1024}$, the present extent, as I understand, of his truly laborious and curious calculations. An Introduction is intended to explain and show examples, of all the various uses of the Tables.

The footnote this time reads:

* I beg here, to correct an inaccuracy and over-sight which occurred last month, in the hasty and very compressed abstract which. for want of room, J was obliged to give in p. 345, of the Labours of the Mathematical Class of the Institute of France; wherein the name of Mr. Farey follows that of M. Laplace, by mistake, instead of the name of *M. Cauchy*,whom the Secretary to the Institute mentions, as having given a general demonstration of the curious property of vulgar Fractions, which Mr. Furey had published in the Philosophical Magazine, Mr. Farey's late Communication to me on the subject, which

TABLE 2.15.    Classification of Farey Papers Cited by Royal Society

| | |
|---|---|
| Geology | 29 |
| Music | 14 |
| Mathematics | 11 |
| Other | 6 |

TABLE 2.16.    Classification of Farey Papers Cited by Hugh Torrens

| | |
|---|---|
| Geology | 114 |
| Music | 63 |
| Farming | 13 |
| Mathematics | 12 |
| Astronomy | 10 |
| Mensuration | 6 |
| Other | 37 |

I have given in.a Note in the page quoted, correctly refers to what had appeared in *"Analyse des Travaux,"* &ce. respect to himself, and the error mentioned, lay entirely with me:-

Beside all of the above Volume 49 of *The Philosophical Magazine and Journal* also contained two very long papers on musical scales by Farey.

### 2.24. Farey's Publications

Table 2.15 is a classification of the papers by John Farey, Sr., listed in the Catalogue of Scientific papers (1800-1863), Volume II, published by the Royal Society of London in 1868 .

Table 2.16 is a classification of Farey's papers in a more complete bibliography compiled by Torrens [**72**].

The fact that Farey picked up on the curious property in Goodwyn's table may be surprising if we think of Farey as a geologist but it isn't so surprising when we recall that Farey researched and published tracts on the mathematics of music. But once again, Farey gets accused of rediscovery.

> Thus the subject stood until 1807, when Mr. John Farey, sen., re-discovered the numbers of Mercator (who never published his mode of deriving them) and showed how they are naturally produced. [12]

The mathematics of music was a popular subject of both writing and debate in the 18th and 19th century. Even Euler [69] weighed in on the subject. The relationship between music and mathematics in the period has been thoroughly explored by Fauvel [73] as well as many others.

In [205] Rasch describes in detail a use of the Farey sequence to analyze music. In discussing the origin of the Farey sequence Rasch says "The series is named after the British geologist John Farey (1977-1826). (Farey also wrote on tuning and temperment but did not apply the series he invented to musical problems.)"

In connecting the primes to music, Farey would seem to have predated du Satory [64] by about 200 years. The reference to page 385 is a pointer to the above curious properties paper so Farey clearly does connect the Farey fractions to his work on music theory. He also seems to understand that the primes are not an arithmetical series.

While Farey may well have recognized that the primes weren't an arithmetic series, this did not stop him from trying to come up with an algorithmic description of the sequence, to the consternation of his readers, and even himself. The following is one such sniff:

> The difference between a man of *real* science, and one who has the ambition to be thought so, is very great. The first seeks to render difficult subjects, perspicuous and clear. The other, on the contrary, envelops even the most simple ideas in the mysterious garb of *hard words* and *scientific jargon*. If Mr. Farey be of the first of those two classes, I should recommend to him to simplify and amend his tables.[222]

TABLE 2.17.   Farey Scale

$$\frac{0}{1}\ \frac{1}{5}\ \frac{1}{4}\ \frac{1}{3}\ \frac{2}{5}\ \frac{1}{2}\ \frac{3}{5}\ \frac{2}{3}\ \frac{3}{4}\ \frac{4}{5}\ \frac{1}{1}\ \frac{5}{4}\ \frac{4}{3}\ \frac{3}{2}\ \frac{5}{3}\ \frac{2}{1}\ \frac{5}{2}\ \frac{3}{1}\ \frac{4}{1}\ \frac{5}{1}\ \frac{1}{0}$$

- A C F A C E F G A c e f g a c e g c e -

$$\frac{1}{1}\quad \frac{5}{4}\quad \frac{4}{3}\quad \frac{3}{2}\quad \frac{5}{3}\quad \frac{2}{1}\qquad \frac{1}{1}\quad \frac{4}{5}\quad \frac{3}{4}\quad \frac{2}{3}\quad \frac{3}{5}\quad \frac{1}{2}$$

c   e   f   g   a   c     c   A   G   F   E   C

In a footnote on page 25 of his book *Introduction to the Psychology of Music*, Révész says

> I should like to call attention here to a mathematical construction of the scale the merit of which seems to be that the major and minor scales are both derived through one and the same principle. This is the Farey 'Series' (1816), which was rigorously proved by Cauchy. Farey considered the series of all non-reducible fractions capable of numerical representation as not greater than an arbitrary number 'n'. If we collocate these fractions in order of value, we obtain the convergent Farey series $F(n)$. For example the series $F(5)$ is as follows: If starting from the middle fraction $1/1$, we identify this with the note c, then the first five fractions to the right represent the six notes of the scale that combine to form the most consonant intervals. In the same way, reading to the left we obtain the six notes of the scale This method gives us a pentatonic scale with the major and minor third. The second and seventh are missing. [**206**]

A little further the above-mentioned paper Rasch says the following:

> We are not the first to use Farey series in musical problems. The delimiting of interval sets with the principles of the Farey series has already been applied by Van der Pol (1946) and Van Eck (1981, pp. 131-36). Other applications in the theory of music can be found in work by Regener (1973) and Rasch (1985). [**205**]

Farey is thus linked to another empirical wanderer, Balthazar van der Pol, not by way of number theory but by way of music theory. And, in the finest tradition of a walk through the valley with James Burke, they both connect to the Riemann hypothesis [**231**]. Of van der Pol's work Mary Cartwright says the following:

> Of van der Pol's papers on the theory of numbers [**231**] is perhaps the best known. In it he combined his knowledge of radio technology and number theory to advantage. In order to investigate the behaviour of the Riemann zeta-function $\zeta$ on the line Re s $= 1/2$ he derives a formula ... the saw-tooth [part of which] was cut on the circumference of a paper disk and a beam of light was projected past the teeth on to a photocell. The electric current so produced eventually yielded a record, rather like an anemometer trace, of the modulus of $(1/2 + it)$, from which the first 73 zeros could be read off with decreasing accuracy for increasing values of t. The branch of number theory, however, which lay closest to his heart was the theory and applications of theta-functions. His published work on this subject is contained in four papers; mention should also be made of his highly individual 'Lectures on a modern unified approach to elliptic functions and elliptic integrals' (mimeographed notes) given at Cornell University in 1958. [**33**]

We will find in the upcoming chapter on applications that there are patents being issued today that harness the properties of the mediant and the Farey sequence. I wonder if the work of these two musical experimenters might not be prior art for a patent recently granted to Stephen Wolfram, the author of Mathematica and *A New Science*[**239**]. Wolfram's patent, US 7,560,636, issued on July 14, 2009, is titled "Method and System for Generating Signaling Sequences" [**240**]. The patent is about the conversion of raw tone sequences into tone sequences that follow a set of aesthetic music principles one of which is "A mediant can be followed by VI, IV, II and V." Rules of this kind were precisely the sort of thing that Farey and his friends, along with many other seekers of the time, were continually and contentiously disputing.

To his credit Wolfram does cite a lot of prior art, but nothing earlier than 1993. Patents based on computerized techniques rarely cite prior art before the invention of the computer itself and yet, it is the technique and not its computer embodiment that is the focus of protection for these patents. If one set out to challenge Wolfram's patent, there's a high likelihood that prior art would be found in *The Ladies Diary*, *The Harmonicon* and *The American Journal of Science, and Arts.*

## 2.25. History's Grudge Against John Farey, Sr.

With Farey's note in our hands, we can see that he didn't try to take credit for discovering the mediant property nor did he try to extend his observation to the generality of a mathematical theorem. And there certainly is nothing in his note that one could possibly call an attempt at a theorem. He says "I am not acquainted, whether this curious property of vulgar fractions has been before pointed out." Nevertheless, so many years later, and even now, Farey is taken to task for poaching Haros' or Goodwyn's discovery.

For example, regarding Farey, one of the most renowned mathematicians of the $20^{th}$ century, G.H. Hardy, is widely and repeatedly quoted in the literature as follows:

> ... Farey is immortal because he failed to understand a theorem which Haros had proved perfectly fourteen years before ...

from [123]and further from [124]

> The history of the 'Farey series' is very curious. Theorems 28 and 29 seem to have been stated and proved first by Haros in 1802; see Dickson, History, i. 156. Farey did not publish anything on the subject until 1816, when he stated Theorem 29 in a note in the Philosophical Magazine. He gave no proof, and it is unlikely that he had found one, since he seems to have been at the best an indifferent mathematician.

Cauchy, however, saw Farey's statement, and supplied the proof (*Exercices de mathématiques, i. 114-16). Mathematicians generally have followed Cauchy's example in attributing the results to Farey, and the series will no doubt continue to bear his name.*

It is not farfetched to consider that it is just a bit disingenuous for a Cambridge don of superstar status in the mathematics community to take a British mineral surveyor – who simply happened to call attention to a curious property of a table of numbers privately published in 1816 – to task for not being familiar with mathematical tables of unknown circulation, printed up during the French Revolution by a reserved French mathematician working in an arcane government bureau some fourteen years earlier, when he happened to call attention to a curious property of a table of numbers privately published in 1816.

Indeed, the libraries that held Haros' tables and the journal in which Haros' paper appeared were undoubtedly more accessible to Cauchy than to Farey, so it's not clear why Hardy works to shift blame for misnaming the series from Cauchy to Farey. Haros' work was presented to the French Academy of Science by de Prony and Legendre. Cauchy surely knew them.

Finally, Hardy must have been aware that in the history of mathematics there are many instances of when theories are rediscovered without the benefit of full citation, just as there are many examples in the field of mathematics of Stigler's Law of Eponymy: No scientific discovery is named after its original discoverer.

Farey's geological work was on a par with the work of William Smith, who is generally acknowledged as the father of geology. Farey and Smith were friends and geological colleagues. Farey is mentioned in Simon Winchester's recent book, *The Map That Changed the World: William Smith and the Birth of Modern Geology* [238]. Farey's treatise and map of Derbyshire [70] were respected during his lifetime and are still respected now.

Farey was an active and wide-ranging contributor to the mathematical literature of the age, including *Philosophical Magazine and Journal* and *The Ladies Diary*. In fact, in the same volume where Farey's note appears about a curious property of Goodwyn's tables, we also find articles by Ampere, Cayley, Laplace and de Prony.

And while his mathematical studies of musical intervals and temperament were quixotic they were certainly not atypical of scientific investigations of the age.

After rereading Hardy's *Apology*, particularly what Hardy has to say about pure and applied mathematics, I have come to a calmer interpretation of Hardy's words about Farey. I believe that the fact that Hardy refers to Farey as a mathematician, however indifferent, amounts at least to a form of grudging praise.

The following is found in the book, *The Mathematical Practitioners of Hanoverian England, 1714-1840*:

> A pupil of the geologist William Smith, Farey calls himself a mineral-surveyor and in 1811 refers to a map dealing with the geological structure of Derbyshire which he had begun in 1807.
>
> He contributed the article on trigonometrical survey to Rees's *Cyclopaedia* and also wrote on the barometrical measurement of heights. In 1812 he wrote the *Memoir* on Derbyshire for the Board of Agriculture. In 1814 Alexander Jamison wrote: 'Mr. Farey, a very ingenious mathematician in London, has lately invented an instrument by which to describe ellipses of any curvature. This instrument may be had at Harris's the opticians.' This was Farey's ellipsograph, described in the *Transactions of the Society of Arts*. Examples survive. [**226**]

It was in fact his son, John Farey, Jr., who invented the ellipsograph.

CHAPTER 3

# The Table Makers

The construction of mathematical tables is largely a lost art. The cross-checking techniques for building tables with crowds of human computers have gone the way of techniques for inverting matrices using mechanical calculators. While we can build accurate, error-free mathematical tables using Maple or Mathematica, today's mathematics major would react to a table with astonishment: "Why would you ever do that?" We save the program that does the computation not the table and run it whenever we need to.

Besides losing a technique for producing useful numbers, we have also lost the aesthetics of mathematical tables and the contribution that intelligently-presented tables can make to experimentation and discovery. With respect to Farey's observation of the mediant property in Goodwyn's tables, Glaisher says the following:

> It seems curious that so elementary and remarkable a property of fractions should not have been discovered until 1816; but supposing the discovery to be due to Mr. Goodwyn and Mr. Farey, an explanation might be afforded by the fact that the 'Tabular Series' is probably the earliest Table of the kind, and that the property would not be likely to present itself to any one who had not arranged a complete series of proper fractions having denominators less than a given number in order of magnitude.[94]

Figures 3.1, 3.2, and 3.3 are specimens from pre-Goodwyn tables that illustrate Glaisher's point. All three tables include the Farey sequence of fractions but displayed in a manner that hinders the discovery of the mediant property.

115

# DENOMINATORS.

| NUMERATORS. | 2 | 3 | 4 | 5 |
|---|---|---|---|---|
| 1 | ,5 | ,3̇ | ,25 | ,2 |
| 2 |  | ,6̇ | ,5 | ,4 |
| 3 |  |  | ,75 | ,6 |
| 4 |  |  |  | ,8 |
| 5 |  |  |  |  |
| 6 |  |  |  |  |
| 7 |  |  |  |  |
| 8 |  |  |  |  |

| NUMERATORS. | 10 | 11 | 12 | 13 |
|---|---|---|---|---|
| 1 | ,1 | ,0̇9̇ | ,08̇3̇ | ,0̇76923̇ |
| 2 | ,2 | ,1̇8̇ | ,16̇ | ,1̇53846̇ |
| 3 | ,3 | ,2̇7̇ | ,25̇ | ,2̇30769̇ |
| 4 | ,4 | ,3̇6̇ | ,3̇ | ,3̇07692̇ |
| 5 | ,5 | ,4̇5̇ | ,416̇ | ,3̇84615̇ |
| 6 | ,6 | ,5̇4̇ | ,5̇ | ,4̇61538̇ |
| 7 | ,7 | ,6̇3̇ | ,583̇ | ,5̇38461̇ |
| 8 | ,8 | ,7̇2̇ | ,6̇ | ,6̇15384̇ |
| 9 | ,9 | ,8̇1̇ | ,75 | ,6̇92307̇ |
| 10 |  | ,9̇0̇ | ,83̇ | ,7̇69230̇ |
| 11 |  |  | ,916̇ | ,8̇46153̇ |
| 12 |  |  |  | ,9̇23076̇ |

FIGURE 3.1. John Marsh, *Decimal Arithmetic Made Perfect*, 1742

| Fractions ordinaires. | FRACTIONS DÉCIMALES. | Fractions ordinaires. | FRACTIONS DÉCIMALES. |
|---|---|---|---|
| 6/19 | 0,315789 | 11/13 | 0,846153 |
| 7/8 | 0,875000 | 11/14 | 0,785714 |
| 7/9 | 0,777777 | 11/15 | 0,733333 |
| 7/10 | 0,700000 | 11/16 | 0,687500 |
| 7/11 | 0,636363 | 11/17 | 0,637059 |
| 7/12 | 0,583333 | 11/18 | 0,611111 |
| 7/13 | 0,538461 | 11/19 | 0,578947 |
| 7/15 | 0,466666 | 11/20 | 0,550000 |
| 7/16 | 0,437500 | 12/13 | 0,923076 |
| 7/17 | 0,411765 | 12/17 | 0,705882 |
| 7/18 | 0,388888 | 12/19 | 0,631579 |
| 7/19 | 0,368421 | 13/14 | 0,928571 |
| 7/20 | 0,350000 | 13/15 | 0,866666 |
| 8/9 | 0,888888 | 13/16 | 0,812500 |
| 8/11 | 0,727272 | 13/17 | 0,764706 |
| 8/13 | 0,615384 | 13/18 | 0,722222 |
| 8/15 | 0,533333 | 13/19 | 0,684211 |
| 8/17 | 0,470588 | 13/20 | 0,650000 |
| 8/19 | 0,421053 | 14/15 | 0,933333 |
| 9/10 | 0,900000 | 14/17 | 0,823529 |
| 9/11 | 0,818181 | 14/19 | 0,736842 |
| 9/13 | 0,692307 | 15/16 | 0,937500 |
| 9/14 | 0,642857 | 15/17 | 0,882353 |
| 9/16 | 0,562500 | 15/19 | 0,789474 |
| 9/17 | 0,529412 | 15/20 | 0,750000 |
| 9/19 | 0,473684 | 16/17 | 0,941176 |
| 9/20 | 0,450000 | 16/19 | 0,842105 |
| 10/11 | 0,909090 | 17/18 | 0,944444 |
| 10/13 | 0,769230 | 17/19 | 0,894737 |
| 10/17 | 0,588235 | 17/20 | 0,850000 |
| 10/19 | 0,526316 | 18/19 | 0,947368 |
| 11/12 | 0,833333 | 19/20 | 0,950000 |

FIGURE 3.2. Claude Collignon, *Découverte D'Etalons Justes*, 1788

# Table.

*De Réduction des fractions vulgaires les plus en usage, en parties décimales.*

| Fraction | Valeur | Fraction | Valeur | Fraction | Valeur | Fraction | Valeur |
|---|---|---|---|---|---|---|---|
| 1/2 | 0,50000 | 3/4 | 0,75000 | 7/8 | 0,87500 | 11/13 | 0,84615 |
| 1/3 | 0,33333 | 3/5 | 0,60000 | 7/9 | 0,77777 | 11/16 | 0,68750 |
| 1/4 | 0,25000 | 3/7 | 0,42857 | 7/12 | 0,58333 | 11/19 | 0,57894 |
| 1/5 | 0,20000 | 3/8 | 0,37500 | 7/12 | 0,46666 | 11/20 | 0,55000 |
| 1/6 | 0,16666 | 3/6 | 0,50000 | 7/16 | 0,43750 | | |
| 1/7 | 0,14285 | 3/16 | 0,17647 | 7/17 | 0,38888 | | |
| 1/8 | 0,12500 | 3/20 | 0,15000 | 7/20 | 0,35000 | | |
| 1/9 | 0,11111 | | | | | | |
| 1/10 | 0,10000 | | | | | 12/13 | 0,92307 |
| 1/11 | 0,09090 | 4/5 | 0,80000 | | | 12/17 | 0,70588 |
| 1/12 | 0,08333 | 4/7 | 0,57142 | 8/9 | 0,88888 | 12/19 | 0,63157 |
| 1/13 | 0,07692 | 4/9 | 0,44440 | 8/11 | 0,72727 | | |
| 1/14 | 0,07142 | 4/11 | 0,36363 | 8/13 | 0,61538 | | |
| 1/15 | 0,06666 | 4/15 | 0,26666 | 8/15 | 0,53333 | | |
| 1/16 | 0,06250 | 4/19 | 0,21052 | 8/17 | 0,47058 | | |
| 1/17 | 0,05882 | | | 8/19 | 0,42105 | 13/16 | 0,81250 |
| 1/18 | 0,05555 | | | | | 13/18 | 0,72222 |
| 1/19 | 0,05263 | | | | | 13/20 | 0,65000 |
| 1/20 | 0,05000 | 5/6 | 0,83333 | | | | |
| | | 5/8 | 0,62500 | 9/10 | 0,90000 | | |
| | | 5/9 | 0,55555 | 9/11 | 0,81818 | | |
| 2/3 | 0,66666 | 5/12 | 0,41666 | 9/16 | 0,56250 | | |
| 2/5 | 0,40000 | 5/16 | 0,31250 | 9/20 | 0,45000 | | |
| 2/7 | 0,28571 | | | | | 14/16 | 0,875 |
| 2/9 | 0,22222 | | | | | 15/16 | 0,93750 |
| 2/11 | 0,18181 | 6/7 | 0,85714 | 10/11 | 0,90909 | 16/17 | 0,94117 |
| 2/13 | 0,15384 | 6/11 | 0,54545 | 10/13 | 0,76923 | 17/18 | 0,94444 |
| 2/15 | 0,13333 | 6/13 | 0,46153 | 10/17 | 0,58823 | 18/19 | 0,94736 |
| 2/17 | 0,11764 | 6/17 | 0,35294 | 10/19 | 0,52631 | 19/20 | 0,95000 |
| 2/19 | 0,10526 | 6/19 | 0,31578 | | | | |

FIGURE 3.3. C. Lewal, *Cours D'Arithmétique Décimale*, 1798

The only table other than Goodwyn's mentioned in histories of mathematical tables that might have sparked Farey's observation is Wucherer's table of decimal fractions for vulgar fractions with numerator and denominator less than 50 [**242**]. Wucherer's table is mentioned in the 1873 British Association for the Advancement of Science report on mathematical tables and in De Morgan's Tables entry in the supplement to the Penny Cyclopedia of 1851 [**53**]. De Morgan says he didn't actually lay eyes on the table whereas Glaisher's committee did. The Glaisher report describes Wucherer's fractions as follows:

> ... arranged according to denominators; so that all having the same denominator are given together; thus the order is
> $$\cdots \frac{1}{17}, \frac{2}{17}, \frac{3}{17}, \cdots \frac{16}{17} \frac{1}{18} \frac{5}{18}, \cdots,$$
> the arguments being only given in their lowest terms. After $\frac{48}{49}$ the system is changed, and the decimals are given for vulgar fractions whose numerators are less than 11 only; thus we have
> $$\frac{1}{50} \frac{2}{50} \frac{3}{50}, \cdots, \frac{10}{50} \frac{1}{51} \frac{2}{51}, \cdots$$
> as consecutive arguments (the arguments not being necessarily in their lowest terms); and the denominators proceed from 50 to 999.[**92**]

Just as with the specimens pictured above, it is highly unlikely that one is going to find the mediant property in Wucherer's fractions. The discovery aspect of tables has as much if not more to do with the visual layout of the table than with the accuracy of the values in the table.

We know the mediant was baked into Haros' table. We don't know how Goodwyn came to the layout in his 1816 table, the one that Farey saw, but as he was such a stickler for accuracy it's hard to imagine that he didn't use the mediant computation to ensure that this table was complete. He knew all about the mediant when he constructed his 1823 table because he includes a description of it in the table itself ... without credit to Haros, Farey or Cauchy.

## 3.1. Archibald's *Mathematical Table Makers*

A quirky yet scholarly monograph entitled *Mathematical Table Makers* by R. C. Archibald appeared in the Scripta Mathematica series published by the Mathematics Department at Yeshiva University [5]. The subtitle of the book is "Portraits, Paintings, Busts, Monuments – Bio-Bibliographical Notes." Archibald sought to make a list of individuals that 1) made mathematical tables **and** 2) for whom at least one picture could be found. For each person that cleared both hurdles, Archibald cites both the tables the person authored and the pictures that are available ... and also selected publications. Sadly neither Charles Haros or Henry Goodwyn made the cut.

TABLE 3.1. Archibald's *Mathematical Table Makers*

| Name | Years | Tables |
|---|---|---|
| John Robinson Airey | 1868 – 1937 | 49 |
| Ernst Emil Ferdinand Anding | 1860 – 1945 | 1 |
| Charles Babbage | 1792 – 1871 | 2 |
| Julius Bauschinger | 1860 – 1934 | 2 |
| Ernst Emil Hugo Becker | 1843 – 1912 | 1 |
| Friedrich Wilhelm Bessel | 1784 – 1846 | 4 |
| David Bierens de Haan | 1822 – 1895 | 3 |
| Jean Charles Borda | 1733 – 1799 | 1 |
| Ernest William Brown | 1866 – 1938 | 2 |
| Joost Bürgi | 1552 – 1632 | 1 |
| Carl Burrau | 1867 – 1947 | 1 |
| Berthold Cohn | 1870 – 1930 | 1 |
| Leslie John Comrie | 1893 – 1950 | 15 |
| Alan Joseph Champneys Cunningham | 1842 – 1928 | 52 |
| Johann Martin Zacharias Dase | 1824 – 1861 | 5 |
| Harold Thayer Davis | 1892 – 1974 | 5 |
| Leonard Eugene Dickson | 1874 – 1954 | 8 |
| Herbert Bristol Dwight | 1885 – 1975 | 9 |
| James Glaisher | 1809 – 1903 | 3 |

TABLE 3.2. Archibald's *Mathematical Table Makers (cont.)*

| Name | Years | Tables |
|---|---|---|
| James Whitbread Lee Glaisher | 1848 – 1928 | 48 |
| Józef Maria Höene-Wronski | 1788 – 1853 | 1 |
| Ernst Reinhold Eduard Hoppe | 1816 – 1900 | 1 |
| Charles Hutton | 1737 – 1823 | 5 |
| Carl Gustav Jacob Jacobi | 1804 – 1851 | 4 |
| Johann Kepler | 1572 – 1630 | 2 |
| Maurice Borisovich Kraïtchik | 1882 – 1957 | 11 |
| Joseph Jérôme le François de LaLande | 1732 – 1807 | 1 |
| Adrien-Marie Legendre | 1752 – 1833 | 15 |
| Derrick Henry Lehmer | 1905 – 1991 | 14 |
| Derrick Norman Lehmer | 1867 – 1938 | 3 |
| Alfred Lodge | 1854 – 1937 | 9 |
| Wilhelm Oswald Lohse | 1845 – 1915 | 1 |
| Eugen Cornelius Joseph von Lommel | 1837 – 1899 | 2 |
| Arnold Noah Lowan | 1898 – 1957 | 36 |
| Andreǐ Andreevich Markov | 1856 – 1922 | 3 |
| Artemas Martin | 1835 – 1918 | 8 |
| Jeffrey Charles Percy Miller | 1906 – 1981 | 10 |
| John Napier | 1550 – 1617 | 2 |
| Niels Nielsen | 1865 – 1931 | 5 |
| Karl Pearson | 1857 – 1936 | 16 |
| Benjamin Osgood Peirce | 1854 – 1914 | 3 |
| Johann Theodor Peters | 1869 – 1941 | 24 |
| Dominique François Rivard | 1697 – 1778 | 1 |
| Edward Sang | 1805 – 1890 | 4 |
| Abraham Sharp | 1651 – 1742 | 1 |
| William Fleetwood Sheppard | 1863 – 1936 | 9 |
| Simon Stevin | 1548 – 1620 | 2 |
| Thomas Jan Stieltjes | 1856 – 1894 | 3 |
| Axel Henrik Hjalmar Tallqvist | 1870 – 1958 | 10 |
| Alexander John Thompson | 1885 – ???? | 3 |
| Herbert Hall Turner | 1861 – 1930 | 2 |
| Horace Schudder Uhler | 1872 – 1956 | 8 |
| François Viète | 1540 – 1603 | 1 |

### 3.2. Lehmer's *Guide to the Tables in the Theory of Numbers*

While neither of our table makers made it onto Archibald's list, Goodwyn did make it into another scholarly and far more exhaustive catalog of mathematical tables of a particular type, D.H.Lehmer's 1941 *Guide to the Tables in the Theory of Numbers*. Archibald was the chairman of the Committee on Mathematical Tables and Aids to Computation that produced this catalog and wrote the Foreword to the report.

A mathematical table is usually thought of as place to go to get a value, the arctangent of $\frac{4}{47}$ for example. This use of a table of numbers is what Haros had in mind when he constructed his rational-to-decimal tables and this use of a mathematical table is certainly what the many people that produced tables of logarithms had in mind.

And yet there is a wholly orthogonal use of mathematical tables that is described very well by Lehmer in the Introduction:

> The theory of numbers is a peculiar subject, being at once a purely deductive and a largely experimental science. Nearly every classical theorem of importance (proved or unproved) has been discovered by experiment, and it is safe to say that man will never cease to experiment with numbers. The results of a great many experiments have been recorded in the form of tables, a large number of which have been published. The theory suggested by these experiments, when once established, has often made desirable the production of further tables of a more fundamental sort, either to facilitate the application of the theory or to make possible further experiments. [**164**]

Unlike the truly experimental sciences however, when a theoretical result is published it is rare that any mention is made of the computations that led to the insight.

There is another quote in Lehmer's Introduction that is in harmony with our efforts in this monograph:

Another peculiarity of the theory of numbers is the fact that many of its devotees are not professional mathematicians but amateurs with widely varying familiarity with the terminology and the symbolism of the subject. In describing tables dealing with those subjects most apt to attract the amateur, some care has been taken to minimize technical nomenclature and notation, and to explain the terminology actually used, while for subjects of the more advanced type no attempt has been made to explain anything except the contents of the table, since no one unfamiliar with the rudiments of the subject would have any use for such a table.

That someone of the stature of D.H. Lehmer would not only acknowledge the existence and contributions of amateurs but would take time and effort to connect his work to them stands in sharp contract to Hardy's haughty dismissal of Farey's note. It is undoubtedly true that Lehmer had the full support of his chairman in this regard for Archibald, too, was an enduring champion of including amateurs as well as professionals in mathematical exploration and discovery.

Back in the 1920's Archibald's inclusive view on mathematics research had resulted in his crossing horns with Oswald Veblen, then the president of the American Mathematical Society. Veblen's view was that the AMS should be an exclusive organization "that admitted only top scholars to membership." [**118**] It was Veblen that had argued against the establishment of the Fields medal at the 1932 International Congress of Mathematicians in Zürich. Evidentally, Veblen wanted to distinguish the elite from the poseurs but having done that he was opposed to distinguishing one elite from another. In spite of his elitist airs Veblen came to the defense of a consulting mathematician. Thomas Grönwall, who was being shunned by the AMS community [**96**]. Even on casual perusal, a stark realization springs to one's mind about the sheer amount of labor that went into the Lehmer report. He 'tabulates' hundreds of tables, together with comparative details, such as digits of accuracy and range of applicability. Bibliographic references are given for each table as well as bibliographic references for corrections in each table. Efforts to reach out to amateurs as well as professionals of modest means include listing libraries in which the referenced tables can be found and an 45-page annex that lists known errata.

**VEGA $1_1$, $1_2$, [$e_1$].**

| N | factors |  | N | factors |
|---|---------|--|---|---------|
| 27293 | $7 \cdot 7 \cdot 557$ |  | 82943 | $7 \cdot 17 \cdot 17 \cdot 41$ |
| 33293 | $13 \cdot 13 \cdot 197$ |  | 90983 | $37 \cdot 2459$ |
| 41779 | $41 \cdot 1019$ |  | 93137 | $11 \cdot 8467$ |
| 55403 | $17 \cdot 3259$ |  | 95017 | $13 \cdot 7309$ |
| 55517 | $7 \cdot 7 \cdot 11 \cdot 103$ |  | 95623 | $11 \cdot 8693$ |
| 57103 | $17 \cdot 3359$ |  |  |  |

(CUNNINGHAM 41, p. 27)

**$1_1$, $1_2$, [$f_1$].**

*delete* 173279, *insert* 177347

(CHERNAC 1; correction of the corresponding table in Vega's *Logarithmisch-trigonometrische Tafeln*, v. 2, Leipzig, 1797, reprinted in VEGA $1_1$, $1_2$.)

FIGURE 3.4. Table of Errata from Lehmer Report

Much like the Royal Society's mathematical table effort described in the chapter on E.H. Neville, the U.S. National Research Council's Committee on Mathematical Tables and Aids to Computation had multiple subcommittees and plans for the publication of mathematical tables, just about all of which were swept aside by the arrival of the computer.

## 3.3. Tables of Tables

3.3 and 3.4 list some catalogs and compilations of tables like those by Archibald and Lehmer described above. The numbers under the *Entries* heading only give a rough indication of the size of the work; they are truly apples and oranges. There is not a consensus among compilers of tables and catalogs of mathematical tables as to what is and is not a mathematical table and if it is a mathematical table whether or not it is sufficiently different from another to be its own entry. For example, some authors make a new entry in their catalog for each table in a collection of tables while other compilers use one entry to describe an entire collection of tables. Furthermore, some compilers count multiple appearances of a table, possibly under a different author's name (and possibly without credit to the original author), while other compilers collect all appearances of a table in one catalog entry. The *Year* column contains the year of publication of the last edition of the compilation.

TABLE 3.3. Compilations of Tables

| Year | Author | Title | Entries | Ref. |
|------|--------|-------|---------|------|
| 1690 | Dechales | Tractatus Proemialis de progressu Matheseos et illustribus Mathematicis | | [58] |
| 1742 | J.C. Heilbronner | Historia Matheseos Universæ a mundo condito ad seculum ... | | [133] |
| 1796 | A.G. Kästner | Geschichte der Mathematik | | [146] |
| 1797 | A. Murhard | Bibliotheca Mathematica | | [183] |
| 1803 | J. De La Lande | Bibliographie Astronomique, avec l'histoire de l'Astronomie | | [155] |
| 1811 | C. Hutton | History of Trigonometric Tables | | [137] |
| 1821 | J.B.J. Delambre | Histoire de L'Astronomie Moderne | | [60] |
| 1828 | J.S. Ersch | Litteratur der Mathematik, Natur- und Gewerbs-Kunde mit Inbegriff der Kriegskunst | | [68] |
| 1830 | J. Rogg | Bibliotheca Mathematica sive Critiçus Librorun Mathematicorum ... | | [210] |
| 1842 | A. De Morgan | TABLES in Penny Cyclopedia | | [52] |
| 1851 | A. De Morgan | TABLES in Penny Cyclopedia Supplement | 318 | [53] |
| 1863 | J.G. Poggendorff | Biographisch-literarisches Handwörterbuch zur Geschichte der exacten Wissenschaften ... | | [199] |
| 1868 | A. De Morgan | TABLES in English Cyclopedia | 457 | [54] |
| 1875 | J.W.L. Glaisher | Report of the Committee on Mathematical Tables | 230 | [92] |
| 1875 | A. Cayley | Report of the Committee on Mathematical Tables 52 | | [36] |

TABLE 3.4. Compilations of Tables

| Year | Author | Title | Entries | Ref. |
|------|--------|-------|---------|------|
| 1902 | R. Mehmke | Numerisches Rechnen | | [178] |
| 1909 | R. Mehmke & d'Ocagne | Calculs Numériques | | [177] |
| 1914 | M.A.Bell and J.R. Milne | A Working List of Mathematical Tables | 240 | [13] |
| 1915 | C.G. Knott | Napier Tercentenary Memorial Volume | | [147] |
| 1926 | J. Henderson | Bibliotheca tabularum mathematicarum: Being a descriptive catalogue of mathematical tables. Part I | 410 | [134] |
| 1929 | L.J. Comrie | Mathematical Tables | | [45] |
| 1941 | D.H.Lehmer | Guide to Tables in the Theory of Numbers | 469 | [164] |
| 1944 | R.C. Archibald | Mathematical Tables in Phil. Mag. | 50 | [4] |
| 1946 | F. A. Rigg | Recent advances in mathematical statistics; bibliography of mathematical statistics, 1940-42 | | [209] |
| 1947 | H.T. Davis | Mathematical Tables | | [50] |
| 1949 | H.T. Davis & Vera Fisher | A Bibliography and Index of Mathematical Tables | | [51] |
| 1948 | R.C.Archibald | Mathematical Table Makers | 433 | [5] |
| 1953 | Walther & Unger | Mathematische Zahlentafeln, numerische Untersuchung spezieller Funktionen | | [236] |
| 1953 | Gertrude Blanch & E.C. Yowell | A Guide to Tables on Punched Cards | | [17] |
| 1954 | D.C. Gilles | Bibliographies on Numerical Calculations | | [91] |
| 1955 | K. Schütte | Index Mathematischer Tafelwerke und Tabellen aus allen Gebieten der Naturwissenschaften | $\approx 1,200$ | [215] |
| 1959 | N.M. Burunova | Spravochnik po matematicheskim tablitsam | 553 | [30] |
| 1960 | A.B. Lebedev | A Guide to Mathematical Tables | 3063 | [161] |
| 1962 | A. Fletcher | An Index of Mathematical Tables, Volume I & II | $> 2,000$ | [77], [78] |

The final and most exhaustive catalog is the two-volume work of Fletcher, Miller, Rosenhead and Comrie whose second edition was published at last light in 1962. This is a staggering work whose 994 pages not only describe thousands of tables by hundreds of table makers but also 152 pages of known errors in these tables and a complete bibliography.

I have not found a reference to Haros' table in any of the catalogs I have examined. Even if one of the compilers knew of the table – and its hard to imagine that one person among Delambre, De Morgan, Glaisher or Lehmer didn't – it is unlikely that it would have been included due to its pedestrian and practical nature. The scholarly value of a table rests on the quality of the numbers it contains, not on the insights it may have sparked. And yet it is with respect to this second measure, ultimately in the author's opinion more important than the first, that Haros' modest table of vulgar fractions and their decimal equivalents outscores all the others.

## 3.4. Neville's Tables

The tables that E.H. Neville prepared in the Royal Society Mathematical Tables series got high marks for their presentation over and above their accuracy. Richard Rado said the following of Neville's *The Farey Series of Order 1025*:

> The production of this volume cannot be overpraised. The Cambridge University Press is well up to its highest standard of printing, the arrangement of the pages is pleasing to the eye and seems to the reviewer to incorporate all the latest anti-eyestrain devices. To anybody who is fond of numbers this book is sure to give many hours of delight.[202]

While it is perhaps hard to imagine curling up in front of hearth of glowing embers with a good table of numbers, table contemplation has resulted in some notable contributions to mathematics, both theoretical and applied. The obscure and opaque realms where academic mathematics operates today can in most cases cases be neither visualized nor tabulated, with the result that our considerable pattern recognition abilities aren't given a chance to contribute to the discovery process.

John Wallis was a much-in-demand cryptanalyst as well as a busy mathematician and man of the cloth. Cryptanalysis, particularly of the form practiced by Wallis, is largely a matter of pattern recognition. Fowler muses

> One can, I think, meaningfully ask why Wallis, the skilful detector of patterns, did not explore the relation between the patterns generated by his approximation procedure, his description of the convergents of Brouncker's continued fraction expansion of $4/\pi$, and his solution, with Brouncker, of Pells' equation even when some features common to the first two examples were brought to his attention by Collins.[81]

An excellent overview of the history of mathematical tables, at least up to 1868, is De Morgan's entry under Tables in the 1868 edition of the English

Cyclopedia [54]. This piece was a greatly expanded version of the same entry in the Penny Cyclopedia [52].

D.H. Lehmer updated De Morgan's history to 1941 at least with respect to tables used for number theory investigations in his *Guide to the Tables in the Theory of Numbers* published by the U.S. National Research Council [164]. For those who enjoy numerical tables, the journal *Mathematical Tables and Aids to Computation*, which was published from April, 1943, to October, 1959, will provide hours of pleasant reading. MTAC, as it was commonly designated, cataloged all sorts of tables old and new, published and unpublished, familiar and arcane. The journal was acquired by the Association of Computing Machinery, renamed the Mathematics of Computation and continues to this day. The unpublished tables and much other MTAC ephemera are archived at the Dolph Briscoe Center for American History at the University of Texas at Austin.

### 3.5. *The Farey Series of Order 1025*

On June 30, 1948, the Committee on Calculation of Mathematical Tables of the British Association for the Advancement of Science transferred all of its assets and liabilities to the Committee on Mathematical Tables of the Royal Society and closed its doors [227]. You might have missed the news since that was the day of the first-ever no-hitter during the evening hours. It was thrown by Bob Lemon in the 2-0 win of the Indians over the Tigers in Briggs Stadium.

The assets of the BAASMTC included what was left of a £3,272 bequest of Lt.-Col. A. J. C. Cunningham for the purpose of producing new tables in the Theory of Numbers and a table "completely or almost completely planned, and for which considerable work has been done" of *The Farey Series* $F_{1025}$ compiled and edited by E. H. Neville [49].

Neville was the person that G.H. Hardy had sent to India in 1914 to persuade Ramanujan to come to England. [158].

It had started to dawn on almost everyone that electronic computers would very soon dump printed tables into the dustbin of mathematics history along with slide rules, protractors and Napier's bones. A member of now-ascendent

TABLE 3.5. Tables of Farey Fractions

| Compiler | Year | Order | Ref. | Comment |
|---|---|---|---|---|
| Mr. Flitcon | 1751 | 100 | [79] | |
| Charles Haros | 1802 | 99 | [128] | |
| Henry Goodwyn | 1818 | 100 | [106] | |
| Henry Goodwyn | 1823 | 1000 | [112] | only 0 to 1/10 |
| Achille Brocot | 1862 | 100 | [26] | |
| J.T. Peters | 1922 | 100 | [195] | extends Brocot |
| Earle Buckingham | 1935 | 120 | [29] | 4 missing fractions |
| Ray M. Page | 1942 | 120 | [191] | logs of decimal equivalent |
| Henry Edward Merritt | 1947 | | [180] | |
| E.H. Neville | 1950 | 1025 | [186] | |

Committee on Mathematical Tables (and as it would turn out, the person destined to lay British table making to its eternal rest) was already busy building the Electronic Delay Storage Automatic Calculator (EDSAC) in the University of Cambridge Mathematical Laboratory, Maurice Wilkes.

Nevertheless in 1948, there were still a few of Cunningham's pounds sterling in the treasury and Neville's table in the pipeline so the Committee on Mathematical Tables stiffened its upper lip and in 1950 published Volume 1 of the Royal Society Mathematical Tables, *The Farey Series of Order* 1025, *Displaying Solutions of the Diophantine Equation* $bx - ay = 1$. *Designed and compiled by E.H. Neville.* The volume sold for £5 s5 in England and $18.50 in New York. Two thousand copies were printed and as of December 31, 1951, only 61 had been sold. Cambridge University Press reprinted the table in 1966. This second edition sold for £7.50 in the UK and $27.50 in the US. In 2010 I acquired a used copy of the first edition for $71.95 and a new copy of the second edition for $95.00.

Table 3.5 presents the entire historical arc of printed tables from their origin to their demise. Their history was launched by the response of an amateur to a query in the *Ladies Diary* and ended with a tabular *tour de force* by a professional published by the Royal Society. §8. *Biographical note* of the Introduction is a scholarly thumbnail history of tables of Farey Fractions.

Neville's motivation to build the table of Farey fractions was as part of the work he was doing on the numerical structure of the Farey sequence; results of which he ultimately published in The Proceedings of the London Mathematical Society [185]. He was interested in the overall distribution and density of Farey fractions and thus wanted to use the table holistically in discovery mode not as a means to retrieve individual values; this is a way that mathematical tables are still used today. Based on Lehmer's entry in *Guide to Tables in the Theory of Numbers* Neville had originally intended to use Goodwyn's tables until he discovered that Lehmer's claim that the entry contained Farey fractions up to denominator 1,000, this actually described Goodwyn's intentions and not his accomplishment. Neville takes Lehmer lightly to task for this discrepancy.

The full name of Neville's Farey table [186] is:

THE FAREY SERIES
OF ORDER 1025
Displaying Solutions of the Diophantine Equation
$$bx - ay = 1$$

The book is dedicated "To the memory of Srinivasa Ramanujan."

The front matter consists of a Preface, an Introduction of 10 sections and a "Decimal index showing the estimated and true locations of $n/1000, n = 0(1)500$." There follows 400 pages of the table itself and then three appendices:

The Farey Series of Order 50 with decimal equivalents
The Farey Series of Order 64
The Farey Integer-Series of Order 100

In the first section of the Introduction, Neville immediately tackles the obvious question:

The order 1025 needs a word of explanation. As a matter of fact it was the series of order 1024 that was first completed; the use of a power of 2 facilitates the preparation, the three-figure limit is comfortably exceeded, and the series is neither so small as to challenge an early extension nor so large as to

```
107  231  124  265  141  158  175  192  209  226   243  260  277   17  284  267  250  233  216  199
384  829  445  951  506  567  628  689  750  811   872  933  994   61 1019  958  897  836  775  714
277  598  321  686  365  409  453  497  541  585   629  673  717   44  735  691  647  603  559  515

182  165  148  279  131  245  114  211   97  274   177  257   80  223  143  206  269   63  235  172
653  592  531 1001  470  879  409  757  348  983   635  922  287  800  513  739  965  226  843  617
471  427  383  722  339  634  295  546  251  709   458  665  207  577  370  533  696  163  608  445

281  109  264  155  201  247   46  259  213  167   121  196  271   75  254  179  283  104  237  133
1008 391  947  556  721  886  165  929  764  599   434  703  972  269  911  642 1015  373  850  477
727  282  683  401  520  639  119  670  551  432   313  507  701  194  657  463  732  269  613  344

162  191  220  249  278   29  273  244  215  186   157  285  128  227   99  268  169  239   70  251
581  685  789  893  997  104  979  875  771  667   563 1022  459  814  355  961  606  857  251  900
419  494  569  644  719   75  706  631  556  481   406  737  331  587  256  693  437  618  181  649

181  111  263  152  193  234  275   41  258  217   176  135  229   94  241  147  200  253   53  277  224
649  398  943  545  692  839  986  147  925  778   631  484  821  337  864  527  717  907  190  993  803
468  287  680  393  499  605  711  106  667  561   455  349  592  243  623  380  517  654  137  716  579
```

223

FIGURE 3.5. Bottom Two Rows on Page 223


be unmanageable. ... The series of order 1024 occupied nearly 399 pages; extension to the order 1025 required exactly one page more.


There are 400 table entries on each page, 319, 765 in all. Figure 3.5 is a scan of the bottom two rows of Page 223. The catch-entry in the bottom right corner is the first Farey fraction on the next page.

The fraction $\frac{1}{2}$ is in the center of each Farey sequence. Terms symmetric to $\frac{1}{2}$ add to one and have the same denominator. In each tabular entry the middle number in bold face is the denominator of the Farey fraction for that entry. The top and bottom numbers are the numerators of the symmetric Farey fractions with this denominator.

Unlike the Haros and Goodwyn tables, Neville doesn't give decimal equivalents. This is the reason there is an unexpectedly long introduction. The largest part of the introduction is devoted to a number of different helpful techniques for finding the tabular entry closest to a decimal value as well as how to use the table to, as the title says, solve Diophantine equations of the form $bx - ay = 1$. Additional pages are spent describing the preparation and production of the table including the many accuracy checks that were made. Neville thanks 40 proof readers including Mrs. E.H. Neville, Mr. L.R. Neville, Mr. M.R. Neville and Miss V.A. Neville.

The tenth and final section of the Introduction elaborates on the dedication. As this book is not widely available, the dedication is given here in its entirety:

> To every mathematician of our time, Farey series recall the name of Srinivasa Ramanujan. For me they revive memories of the man himself, who became my friend in the golden twilight of our lost civility, the earlier half of the year 1914. Ramanujan's place in history was determined not when his first letter to Hardy showered fireworks in the Cambridge sky in 1913, for he resisted all immediate efforts to attract him to England, but when in Madras a year later he put into my hands the now famous Notebook and suggested that I should take it and examine it at my leisure. Only in a slight degree was the compliment personal to me; to Ramanujan then Hardy was a name on paper, and Walker and Littlehailes were parts of the governing machine, but I was a human being. Nevertheless, if Hardy could long remember with satisfaction that he could recognize at once what a treasure he had found, I too can be proud, for if I had failed to win the confidence of Ramanujan and his friends, Ramanujan would not have followed me to England. Nor could my failure have been remedied; before the autumn, that is, before any other visitor from Cambridge could have made contact with him, war was raging, and five years were to elapse before communication between India and England was again easy.
>
> Ramanujan's first three months in England were spent in my house in Cambridge, and the friendship which developed ended only with his death.

## 3.6. Reviews of *The Farey Series of Order 1025*

Somewhat surprisingly *The Farey Series of Order 1025* received four reviews in the literature, one by Raymond Archibald in *Mathematical Tables and Other Aids to Computation* [**6**], another by Paul Bateman in the *Bulletin of the*

| 152 | 457 | 305 | 458 | 153 | 460 | 307 | 461 | 154 | 463 | | 309 | 464 | 155 | 466 | 311 | 467 | 156 | 469 | 313 | 470 |
|-----|-----|-----|-----|-----|-----|-----|-----|-----|-----|--|-----|-----|-----|-----|-----|-----|-----|-----|-----|-----|
| 305 | 917 | 612 | 919 | 307 | 923 | 616 | 925 | 309 | 929 | | 620 | 931 | 311 | 935 | 624 | 937 | 313 | 941 | 628 | 943 |
| 153 | 460 | 307 | 461 | 154 | 463 | 309 | 464 | 155 | 466 | | 311 | 467 | 156 | 469 | 313 | 470 | 157 | 472 | 315 | 473 |

FIGURE 3.6. Row with Errors on Page 399

*American Mathematical Society* [11], a third by Richard Rado in *The Mathematical Gazette* [202] and a fourth by Derrick Lehmer in *Mathematical Reviews* [MR0038362 (12,392c)].

The review by Raymond Archibald, a history of mathematics professor at Brown University, is a detailed and elegant thumbnail sketch of the entire history of the Farey sequence that has become the canonical reference for this history. But, as every tough reviewer must, Archibald does find a fault:

> To ensure accuracy in the splendid new volume of the British Mathematical Tables Committee more than 40 people collaborated with Professor Neville. Hence we must wonder why neither they nor the printer notices that on page 399 are ten numbers which should have been in ordinary rather than blackface type. [6]

In his review Rado writes, "To anybody who is fond of numbers this book is sure to give many hours of delight." In the same review, Rado cites two applications of the table. One is solving the Diophantine equation in the title even when $a$ and $b$ aren't in the table. Neville calls the procedure the cascade method and it is covered in the Introduction as well as in [185]. The other application Rado mentions is finding a rational approximation with a prescribed upper bound on the denominator.

On the other hand, in his review, Paul Bateman says:

> Although this work will be a mere curiosity to most mathematicians and will certainly not have widespread use, number-theoretic experimenters will find it of considerable interest. [11]

Lehmer's review in Mathematical Reviews was short and to the point:

> This volume is the first of a new series of British tables spon-
> sored by the Royal Society. The main table gives the numera-
> tors and denominators of all reduced rational numbers $P_n/Q_n$
> with $0 < P_n < Q_n \le 1025$ arranged in increasing order. The
> table contains $1 + \sum_{n=1}^{1025} \phi(n) = 319765$ such fractions and is
> nearly 100 times as extensive as previous tables. The sequence
> of fractions is "folded back on itself" at $\frac{1}{2}$, the numerators of
> $P_n$ and $Q_n - P_n$ being given in the same place with $Q_n$. No
> approximate decimal equivalents are given, but due to the re-
> markably uniform distribution of these values throughout most
> of the table, it is not difficult to locate by page and line num-
> ber the position of a given fraction, especially if use is made
> of the two page decimal index. The introduction illustrates
> the application of the table to such problems as the rational
> approximation to irrationals and the solution of linear Dio-
> phantine equations. There are four appendices. (1) The Farey
> series of order 50. Here decimal equivalents are also given to
> 5D. (2) The Farey series of order 64 on a single page. (3) The
> denominators of the Farey series of order 100. (4) A factor
> table of the first 1049 integers on both sides of a loose card.

## 3.7. Solving Diophantine Equations

While it would further his own research, Neville sold the table project to
the British Association for the Advancement of Science on the basis of the
table's ability to quickly and easily solve a certain class of linear Diophantine
equations.

What Neville refers to as the Haros property of the Farey sequence, that
$bc - ad = 1$ if $\frac{a}{b}$ and $\frac{c}{d}$ are adjacent Farey fractions, enables one to instantly
read out a solution to

$$bx - ay = \pm 1$$

when $a < b \le 1025$ as one adjacent fraction or the other of $\frac{a}{b}$. In fact, sur-
rounding fractions with denominators less than $b$ provide alternative solutions

for which $y < b$ but as Neville notes, one solution is sufficient to generate all the others: "if $p$, $q$ is one solution of $bx - ay = \pm 1$, the general solution is $p + at, q + bt$, for an arbitrary integral value of $t$, positive, zero, or negative, and the value of $t$ such that $0 < q + bt < b$ is discoverable immediately."

For the situation where $b$ exceeds 1025, Neville describes his cascade solution of the linear Diophantine equation. Here's his example using the leading digits of $\pi$ and $e$. Start with the equation

$$314159265359x - 271828182846y = 1.$$

The ratio of the coefficients, 0.865256, is between $\frac{122}{141}$ and $\frac{777}{898}$ in the table so make the substitution

$$x = 777x_1 + 122y_1 \qquad \text{and} \qquad y = 898x_1 + 141y_1$$

to form the equation in $x_1$ and $y_1$

$$40988235(x_1 8y_1) - 15501608y_1 = 1.$$

Go around again. The ratio of the two coefficients, 0.378197, is between $\frac{281}{743}$ and $\frac{340}{899}$ so we make the substitution

$$x_1 - 8y_1 = 340x_2 + 218y_2 \qquad \text{and} \qquad y_1 = 899x_2 + 743y_2$$

to form the equation

$$709(77x_2 - y_2) - 285x_2 = 1$$

which can be solved from Neville's $F_{1025}$ table as

$$709 \cdot 244 - 285 \cdot 607 = 1.$$

We can now unwind the substitutions giving

$$x_2 = 607 \qquad \text{and} \qquad y_2 = 46495$$

and then

$$x_1 = 294003299 \qquad \text{and} \qquad y_1 = 35091478$$

and finally

$$x = 232721723639 \qquad \text{and} \qquad y = 268962860900.$$

In a distant yet clearly audible echo of Haros, the cornerstone of these computations is getting from a decimal value to a best approximating vulgar fraction. This is exactly the problem that Neville was studying – the structure of the Farey sequence – when he discovered that Goodwyn's table didn't exist and embarked on the subtask of computing the Farey Series of Order 1025.

### 3.8. *Rectangular-Polar Conversion Tables*

Neville was also the designer and compiler of Volume 2 in the Royal Society Mathematical Tables series, *Rectangular-Polar Conversion Tables*. Like *The Farey Series of Order 1025*, work on this table was begun by British Association for the Advancement of Science using the bequest of Lt.-Col. A.J.C Cunningham and finished under the auspices of the Royal Society by Neville. This was the last table he completed; the last section of the Introduction is §14. *Envoi.*

At the very same time as Neville was completing his table of arctangents, another Englishman, John Todd, was completing a table of arctangents for the U.S. National Bureau of Standards. Why in the mid 1950's would anybody care about high-precision tables of the arctangent? Todd gives a hint in the introduction to his table:

> For some time there has been a considerable demand for a table of Arctangents with rational arguments. This function occurs, for instance, as the imaginary part of $\log n(m + in)$ and as the polar angle of the point $(m, n)$. An important use of $\log(m+in)$ is made in the tabulation of the function $\log \Gamma(x + iy)$, which was undertaken simultaneously in England by J. C. P. Miller and S. Johnston and in the United States by the Computation Laboratory of the National Bureau of Standards. This function appears in the asymptotic expansion of Whittaker's confluent hypergeometric function, which satisfies a differential equation of importance in many atomic collision problems, ..., and it is this fact that explains some of the demand for the tables of Arctangents. [**229**]

Neville and Todd used different methods to compute their arctangent tables but they both started with the Farey sequence, or as Neville says "with the Haros property of Farey series," for exactly the same reason that Charles Haros did – the need for a complete list of all irreducible fractions with denominators less than a given value. Neville and Todd reference each other and each checked the values in their table against the other's table during proof.

Neville relied on the identity

$$\arctan z_1 \pm \arctan z_2 = \arctan\left(\frac{z_1 \pm z2}{1 \mp z_1 z_2}\right) \tag{3.8.1}$$

for his approximation whereas Todd used the factorization of arctan that he invented [**228**]

$$\arctan n = \sum f_r \arctan n_r$$

as his approximation. Neville uses the Madhava-Gregory series for arctan

$$\arctan x = \sum_{i=1}^{\infty} -1^{i-1}\frac{x^{2i-1}}{2i-1}$$

to compute selected values of the arctan to feed into the approximation.

Neville illustrates his approach in §2. *The computation of* $\Theta(x, y)$ *in the In-troduction.* This tutorial is a sterling example of the unique quality of the mathematical thinking that underpins table building that we saw above in Haros' published papers. The example also gives a hint as to why the Farey sequence shows up so often in table-building projects. Indeed, it is used by Neville to build his rectangular-polar conversion tables for exactly the same reason that Haros used it 150 years before to build his fraction-decimal conversion tables.

In converting from rectangular coordinates to polar coordinates we need to compute the angle between the x-axis and the line from the origin through the point in rectangular coordinates, $(x, y)$, namely $\arctan(y/x)$. To minimize the published size of the table we only need to include irreducible ratios with the numerator less than the denominator. From these values one can easily compute the angle associated with excluded $(x, y)$ pairs. Therefore, if the table is going to cover $(x, y)$ pairs for which $1 \le y \le x \le 6$ then the table is nothing more than the angles whose tangents are the elements of $F_6$, the Farey sequence of order 6; viz.

$$\frac{0}{1}, \frac{1}{6}, \frac{1}{5}, \frac{1}{4}, \frac{1}{3}, \frac{2}{5}, \frac{1}{2}, \frac{3}{5}, \frac{2}{3}, \frac{3}{4}, \frac{4}{5}, \frac{5}{6}, \frac{1}{1}$$

Neville launches his table-building by immediately harnessing the Haros property of sequential elements $\frac{a}{b}$ and $\frac{c}{d}$ Farey sequence, $bc - ad = 1$, to form an identity for the first difference of the angles with these tangents; to wit,

$$\arctan\left(\frac{a}{b}\right) - \arctan\left(\frac{c}{d}\right) = \arctan\left(\frac{1}{bd + ac}\right) \tag{3.8.2}$$

Neville uses this identity to compute the tangents of the angles that are the first difference angles whose tangents are in $F_6$. For example,

$$\arctan\left(\frac{1}{5}\right) - \arctan\left(\frac{1}{6}\right) = \arctan\left(\frac{1}{5\cdot 6 + 1\cdot 1}\right) = \arctan\left(\frac{1}{31}\right)$$

so $\frac{1}{31}$ is the tangent of the angle that is the difference between the angle whose tangent is $\frac{1}{5}$ and angle whose tangent is $\frac{1}{6}$. Applying this identity to $F_6$ gives him the sequence of tangents

$$\frac{1}{6}, \frac{1}{31}, \frac{1}{21}, \frac{1}{13}, \frac{1}{17}, \frac{1}{12}, \frac{1}{13}, \frac{1}{21}, \frac{1}{18}, \frac{1}{32}, \frac{1}{50}, \frac{1}{11}, \tag{3.8.3}$$

among which there are only 10 different values

$$\frac{1}{50}, \frac{1}{32}, \frac{1}{31}, \frac{1}{21}, \frac{1}{18}, \frac{1}{17}, \frac{1}{13}, \frac{1}{12}, \frac{1}{11}, \frac{1}{6}. \tag{3.8.4}$$

To these he applies his approximation, equation 3.8.1, to get the tangents of the angles that are the difference between the angles with these tangents:

$$\frac{18}{1601}, \frac{1}{993}, \frac{10}{652}, \frac{3}{379}, \frac{1}{307}, \frac{4}{222}, \frac{1}{13}, \frac{1}{157}, \frac{1}{133}, \frac{5}{67}. \tag{3.8.5}$$

At this point he appeals to the Madhava-Gregory expansion of arctan to observe that for all these values $t$ save the last one

$$\frac{t^5}{5} < 10^{-9}$$

so the easily-managed first two terms of the expansion,

$$t - \frac{t^3}{3} \tag{3.8.6}$$

can be used to compute the needed angles to the target 7-place accuracy. For $\frac{5}{67}$ one additional term is needed. This gives Neville the nine angles whose tangents are listed in 3.8.5;

$$0.01124250, 0.00100705, 0.01533622, 0.00791540, 0.00325732, 0.01801607,$$

$$0.00636934, 0.00751866, 0.07448879.$$

Using 3.8.6 again to compute $\arctan\left(\frac{1}{50}\right)$ as $0.01999733$ he can add $0.01124250$ to this to get $\arctan\left(\frac{1}{32}\right)$ and then add $0.00100705$ to get $\arctan\left(\frac{1}{31}\right)$, and so forth, unrolling the second set of first differences. This gives

$$0.01999733, 0.00100705, 0.03123983, 0.03224688, 0.04758310, 0.05549850,$$

$$0.05875582, 0.07677189, 0.08314123, 0.09065989, 0.16514868$$

as the angles whose tangents are in the list 3.8.4, and, where necessary, by reordering and duplication the list 3.8.3. Starting with 0.16514868, the angle whose tangent is $\frac{1}{6}$, we unroll again. Add 0.03224688 to this value to get 0.19739556, the angle whose tangent is $\frac{1}{5}$. Keep adding the first differences to get each of the angles whose tangents are in $F_6$,

$$0.16514868, 0.19739556, 0.03123983, 0.24497866, 0.32175055,$$

$$0.38050637, 0.46364760, 0.54041951,$$

$$0.58800261, 0.64350111, 0.67474094, 0.69473827, 0.78539816.$$

We computed nine rational cubes and one integer fifth power which can be obtained by lookup in existing tables. All the rest was addition and subtraction.

For Haros and Neville and for all table makers in between the goal is always to convert the computation of complex values to the most rudimentary arithmetic operations, the most rudimentary of all being table lookup. This is how squares and square roots, cubes and cube roots were handled. Everybody had a table of powers and roots at hand and these tables being themselves rather rudimentary were of high quality from the beginning. The second most simple arithmetic operations are addition and subtraction. The reduction of the bulk of computations to just these two was the goal of every table builder, which is why first difference roll-ups were a core strategy of so many table projects.

Value logistics as opposed to value computation was one of the most challenging aspects of table-building. Even if the look-up or the addition was performed flawlessly, if the result was put in the wrong place, the rolling differences strategy turned from efficiency maker to table destroyer as the error cascaded from one value to the next. From the very beginning of a table-building project when computing seed and anchor values such as $\arctan\left(\frac{1}{50}\right)$ in the above example, until the very end, when reviewing printer's proofs, the programmer's bête noire – being off by one – not in the value but in the slot number, remained an obsessive worry. In fact I did Neville an injustice in the above description because all of his roll-ups were done from both ends of the sequences with meet-in-the-middle performing a check that there had been no index slips.

While mathematical tables are a thing of the past, the type of mathematical thinking and algorithms behind table construction are alive and well today in the construction of software, when faced with severe space and time constraints. While it takes care and creativity one can buy speed with precision – as witness the fast Fourier transform, for example – and this is exactly what table builders strove for. Their unit of time just happened to be months rather than microseconds.

Aborted and abandoned projects are as much a recurring theme in the history of mathematical table building as is the creation of tables that find wide popularity and daily use. De Prony's project [115] is the poster child and Sang's lost tables are also an oft-cited example. Neville's introduction to the rectangular-polar conversion tables contains a poignant side story that gives one a glimpse of the life of the people doing the actual computing.

> In 1936 I seeded the octant with angles whose sines and cosines are terminating decimals, that is, Pythagorean angles with hypotenuses of the form $2^m 5^n$, and Miss E. J. Ternouth began to calculate, for $x, y$ pairs in the range $1 \leq y \leq x \leq 100$, the differences, in degrees to 10 places, between $\Theta(x, y)$ and the nearest of these angles. The work was tedious, and the routine included no automatic check; Miss Ternouth's results, patiently accumulated and now completely superseded, could not be utilized in any way in the process ultimately found practicable, but her labours deserve mention. [187]

Miss Ternouth is also credited as a contributor to Neville's *Farey Series of Order 1025* described above and is a named author of two of the British Association Mathematical Tables, [194] and [95].

### 3.9. **Reviews** of *Rectangular-Polar Conversion Tables*

Neville's *Rectangular-Polar Conversion Tables* received three reviews in the literature. They included a lengthy review by John Todd, the man in friendly competition at the National Bureau of Standards, in *Mathematical Reviews* [MR0077245 (17,1011a)], a second by John Wrench in *Mathematical Tables and Other Aids to Computation* [241], and a third by Alan Fletcher in *The Mathematical Gazette*.

Todd's review includes a deft description of Neville's approach.

> The new material is mainly the values of $\theta$. It was prepared
> essentially as follows. Since $x_i y_{i+1} - x_{i+1} y_i = 1$ we have arctan
> $f_{i+1} = \arctan f_i + \arctan n_i^{-1}$ where $n_i = x_i x_{i+1} + y_i y_{i+1}$ and
> so arctan $f_i$ could have been built up from an initial value
> and the values of the differences, arctan $n_i^{-1}$. Actually the
> sequence arctan $n_i^{-1}$ was itself first built from its differences.
> More precisely, the sequence $\{n_i\}$ was arranged in order of
> magnitude and without repetitions as $\{m_i\}$ and then arctan
> $m_i^{-1}$ was built up from its differences, arctan $p_i$, where
>
> $$p_i = (m_{i+1} - m_i)/(1 + m_i m_{i+1}).$$
>
> In most cases $p_i < 10^{-7}$ and so arctan $p_i = p_i$ to about
> 20D. Many precautions were taken in the preparation: e.g., the
> sequences were built up from both ends, the building up was
> done for both the measures of the $\theta$, instead of in one case,
> followed by a final multiplication to convert the tabular entry.
> [**76**]

Wrench notes that

> The versatility of the conversion tables is illustrated by their
> application to the calculation of the principal value of $(e + i\gamma)^\pi$
> to about eleven significant figures in both rectangular and polar
> form.

The reviews by both Wrench and Fletcher end with high compliments for
the editor and the publisher. Wrench closes by saying

> The high standard of typographical excellence characteristic
> of the earlier publications of both this committee and their
> predecessors has been maintained. The present set of tables,
> in providing reliable key values for future tabulations, as well
> as very accurate working data, constitutes a valuable addition
> to the growing literature of such numerical information.

and Fletcher by

> The introduction is graced by the compiler's customary distinction of style, and the whole volume by the usual Cambridge excellence of production.

## 3.10. Moritz Stern and Achille Brocot

Seventy-six years after Haros published his table of fraction-to-decimal tables, Achille Brocot published his own [**27**]. Unlike Haros, however, who only intended that his table ease translation between rational and decimal representations of numbers at the time the decimal system was coming on line, Brocot had a very particular application in mind – making the gears in watches. Figure 3.7 is a page from Brocot's table.

If in applying the mediant function one disregards the condition that denominators be less than or equal to a given value as one did in building the Farey sequence and starts with the pair of fractions $\frac{0}{1}$ and $\frac{1}{0}$ rather than $\frac{0}{1}$ and $\frac{1}{1}$ one generates the sequence of Stern-Brocot sequences rather than the sequence of Farey sequences:

$$B_0 = \left(\frac{0}{1}, \frac{1}{0}\right)$$

$$B_1 = \left(\frac{0}{1}, \frac{1}{1}, \frac{1}{0}\right)$$

$$B_2 = \left(\frac{0}{1}, \frac{1}{2}, \frac{1}{1}, \frac{2}{1}, \frac{1}{0}\right)$$

$$B_3 = \left(\frac{0}{1}, \frac{1}{3}, \frac{1}{2}, \frac{2}{3}, \frac{1}{1}, \frac{3}{2}, \frac{2}{1}, \frac{3}{1}, \frac{1}{0}\right)$$

$$B_4 = \left(\frac{0}{1}, \frac{1}{4}, \frac{1}{3}, \frac{2}{5}, \frac{1}{2}, \frac{3}{5}, \frac{2}{3}, \frac{3}{4}, \frac{1}{1}, \frac{4}{3}, \frac{3}{2}, \frac{5}{3}, \frac{2}{1}, \frac{5}{2}, \frac{3}{1}, \frac{4}{1}, \frac{1}{0}\right)$$

Moritz Stern published a paper [**224**] on this sequence in 1858, two years earlier than Brocot in Crelle's Journal, and discussed it in purely mathematical terms. Thus, in this case, at least at the time of this writing, the series is correctly called the Stern-Brocot series.

— **21** —

| Z | N |  | Z | N |  |
|---|---|---|---|---|---|
| 4 | 31 | 0,129 032 258 06 | 9 | 64 | 0,140 625 000 00 |
| 11 | 85 | 129 411 764 71 | 10 | 71 | 140 845 070 42 |
| 7 | 54 | 129 629 629 63 | 11 | 78 | 141 025 641 03 |
| 10 | 77 | 129 870 129 87 | 12 | 85 | 141 176 470 59 |
| 13 | 100 | 130 000 000 00 | 13 | 92 | 141 304 347 83 |
| 3 | 23 | 130 434 782 61 | 14 | 99 | 141 414 141 41 |
| 11 | 84 | 130 952 380 95 | 1 | 7 | 142 857 142 86 |
| 8 | 61 | 131 147 540 98 | 14 | 97 | 144 329 896 91 |
| 13 | 99 | 131 313 131 31 | 13 | 90 | 144 444 444 44 |
| 5 | 38 | 131 578 947 37 | 12 | 83 | 144 578 313 25 |
| 12 | 91 | 131 868 131 87 | 11 | 76 | 144 736 842 11 |
| 7 | 53 | 132 075 471 70 | 10 | 69 | 144 927 536 23 |
| 9 | 68 | 132 352 941 18 | 9 | 62 | 145 161 290 32 |
| 11 | 83 | 132 530 120 48 | 8 | 55 | 145 454 545 45 |
| 13 | 98 | 132 653 061 22 | 7 | 48 | 145 833 333 33 |
| 2 | 15 | 133 333 333 33 | 13 | 89 | 146 067 415 73 |
| 13 | 97 | 134 020 618 56 | 6 | 41 | 146 341 463 41 |
| 11 | 82 | 134 146 341 46 | 11 | 75 | 146 666 666 67 |
| ʼ9 | 67 | 134 328 358 21 | 5 | 34 | 147 058 823 53 |
| 7 | 52 | 134 615 384 62 | 14 | 95 | 147 368 421 05 |
| 12 | 89 | 134 831 460 67 | 9 | 61 | 147 540 983 61 |
| 5 | 37 | 135 135 135 14 | 13 | 88 | 147 727 272 73 |
| 13 | 96 | 135 416 666 67 | 4 | 27 | 148 148 148 15 |
| 8 | 59 | 135 593 220 34 | 11 | 74 | 148 648 648 65 |
| 11 | 81 | 135 802 469 14 | 7 | 47 | 148 936 170 21 |
| 3 | 22 | 136 363 636 36 | 10 | 67 | 149 253 731 34 |
| 13 | 95 | 136 842 105 26 | 13 | 87 | 149 425 287 36 |
| 10 | 73 | 136 986 301 37 | 3 | 20 | 150 000 000 00 |
| 7 | 51 | 137 254 901 96 | 14 | 93 | 150 537 634 41 |
| 11 | 80 | 137 500 000 00 | 11 | 73 | 150 684 931 51 |
| 4 | 29 | 137 931 034 48 | 8 | 53 | 150 943 396 23 |
| 13 | 94 | 138 297 872 34 | 13 | 86 | 151 162 790 70 |
| 9 | 65 | 138 461 538 46 | 5 | 33 | 151 515 151 52 |
| 5 | 36 | 138 888 888 89 | 12 | 79 | 151 898 734 18 |
| 11 | 79 | 139 240 506 33 | 7 | 46 | 152 173 913 04 |
| 6 | 43 | 139 534 883 72 | 9 | 59 | 152 542 372 88 |
| 13 | 93 | 139 784 946 24 | 11 | 72 | 152 777 777 78 |
| 7 | 50 | ·140 000 000 00 | 13 | 85 | 152 941 176 47 |
| 8 | 57 | 140 350 877 19 | 15 | 98 | 153 061 224 49 |
| 9 | 64 | 140 625 000 00 | 2 | 13 | 153 846 153 85 |

FIGURE 3.7. Table from Brocot's Book

Some applications of the Stern-Brocot tree use only the fractions being added at each step, $G_i = B_i/B_{i-1}$, not all of the fractions computed to date.

$$G_0 = \left( \frac{0}{1}, \frac{1}{0} \right)$$

$$G_1 = \left( \frac{1}{1} \right)$$

$$G_2 = \left( \frac{1}{2}, \frac{1}{1}, \frac{2}{1} \right)$$

$$G_3 = \left( \frac{1}{3}, \frac{2}{3}, \frac{3}{2}, \frac{3}{1} \right)$$

$$G_4 = \left( \frac{1}{4}, \frac{2}{5}, \frac{3}{5}, \frac{3}{4}, \frac{4}{3}, \frac{5}{3}, \frac{5}{2}, \frac{4}{1} \right)$$

Just as the Farey sequence finds all fractions between 0 and 1, the Stern-Brocot sequence finds all positive rational numbers; viz. all fractions between 0 and $\infty$. And also, as with the Farey sequence, there are a slew of formulamatic relationships between the elements of the Stern-Brocot tree all driven by the same foundational one: If $\frac{a}{b}$ and $\frac{c}{d}$ are consecutive fractions then

$$cb - ad = 1.$$

In the table-contemplation tradition of John Farey, Pierre Lamothe is credited with observing that for the fractions added at the $i^{th}$ step

$$\sum_{\frac{p}{q} \in G_i} \frac{1}{pq} = 1.$$

Whether somebody will step forward to play Haros to Lamothe's Farey is an open question.

## 3.11. Gears and Rational Approximation

If you've got a shaft turning at $t$ revolutions per second and you want to turn a second shaft at $s$ revolutions per second what you need to do is put a gear with $s_0$ teeth on the first shaft that drives a gear with $t_0$ teeth on the second shaft such that

$$\frac{s}{t} = \frac{s_0}{t_0}$$

If $t$ is 20 and $s$ is 10 then since the teeth mesh one-for-one, a gear with 535 teeth on the first shaft driving a gear with 1070 teeth on the second shaft will cause the second gear to rotate once for every two of its own rotations.

For an n-stage gear train where intermediate gear $i$ has a driven gear of $s_i$ teeth and a driving gear of $t_i$ teeth we need to have

$$\frac{s}{t} = \prod_{i=0}^{n} \frac{s_i}{t_i}$$

We factor $s$ (respectively $t$) and group the factors into $n$ groups. In our example we could use a single intermediate gear with 25 teeth being driven by a 21 tooth gear on the first shaft and 30 teeth driving a 35 tooth gear on the final shaft.

All of this works fine as long as $s$ and $t$ can be factored in a way that produces a physically feasible solution. Physically feasible roughly means that $n$ isn't too big, and the number of teeth on any one gear isn't too large. If $t$ is 23 and $s$ is 191, a motivational example considered by Brocot, one has to construct a physically feasible approximation to $\frac{191}{23}$ rather than a solution, because both numbers are prime. Defining and computing such an approximation is the problem addressed by Brocot.

Stern's interest was in the relationship between the mediant and the Euclidean algorithm. He made the same observation that Wallis had made 175 years earlier that if you keep track of the number of successive times you go left or right as you head toward a number you generate the digits in the continued fraction expansion of the number. Essentially, Stern rediscovered the Parmenides Algorithm. Every fraction appears sooner or later in the Stern-Brocot tree and the path through the tree to particular fraction is just the application of Chuquet's *règle des nombres moyens*. What Chuquet didn't appreciate is that each fraction $f$ along the way is the best approximation among all fractions with numerators less than $f$'s and denominators less than $f$'s. Making gears with lots of teeth is harder than making gears with fewer teeth so making the best possible gear with an upper bound on the number of teeth is obviously of interest to a watchmaker.

Brocot does not cite Chuquet nor does Stern cite Wallis and neither of them cite Lagrange [**154**], who also used continued fraction expansions to compute rational approximations with constraints on the numerators and denominators of the approximants. Perhaps we have another instance of Stigler's law of eponymy. Maybe the Stern-Brocot tree should more accurately be called the Chuquet-Wallis tree.

In the single-gear case, walking the Stern-Brocot tree gives one the definition of an optimal solution; namely, the best single gear pair with an upper bound on both the number of teeth in the driving and driven gears. This property of the walk also produces candidates worthy of consideration for multi-stage gear trains.

Consider a problem posed by Camus in the chapter on gears in *Cours de Mathématique* (translated into English by Hawkins [**132**]):

> 581. To find the number of teeth and leaves of the wheels and pinions of a machine, which being moved by a pinion, placed on the hour wheel, shall cause a wheel to make a revolution in a mean year, supposed to consist of 365 days, 5 hours, 49 minutes. [**32**]

Converting everything to minutes, $t$ is 720 and $s$ is $525,949$. Camus recommends a 4-stage solution with

$$\frac{3 \times 25 \times 23 \times 83}{2 \times 2 \times 7 \times 7} = \frac{196}{143175}$$

at depth 753 in the Stern-Brocot tree. As

$$60\left(525949 - 720\frac{143175}{196}\right) \approx 1.2$$

Camus' solution loses about $1\frac{1}{5}$ second per year.

At depth 745 in the Stern-Brocot tree we find a 2-stage solution

$$\frac{3 \times 3}{11 \times 63} = \frac{27}{19723}$$

which loses about

$$525949 - 720\frac{19723}{27} \approx 2\frac{1}{3}$$

minutes per year.

Figure 3.8 is a page from Section 4 of Page's *14000 Gear Ratios* book [**191**] that shows recommended 2-stage gear combinations to achieve a particular sum.

| N. | FACTORS | N. | FACTORS | N. | FACTORS | N. | FACTORS |
|---|---|---|---|---|---|---|---|
| 3420 | 38 x 90 | 3481 | 59 x 59 | 3552 | 37 x 96 | 3621 | 51 x 71 |
| 3420 | 45 x 76 | 3483 | 43 x 81 | 3552 | 48 x 74 | 3626 | 37 x 98 |
| 3420 | 57 x 60 | 3484 | 52 x 67 | 3555 | 45 x 79 | 3626 | 49 x 74 |
| 3422 | 29 x 118 | 3485 | 41 x 85 | 3560 | 40 x 89 | 3627 | 31 x 117 |
| 3422 | 58 x 59 | 3486 | 42 x 83 | 3564 | 33 x 108 | 3627 | 39 x 93 |
| 3424 | 32 x 107 | 3488 | 32 x 109 | 3564 | 36 x 99 | 3630 | 33 x 110 |
| 3430 | 35 x 98 | 3492 | 36 x 97 | 3564 | 44 x 81 | 3630 | 55 x 66 |
| 3430 | 49 x 70 | 3496 | 38 x 92 | 3564 | 54 x 66 | 3634 | 46 x 79 |
| 3431 | 47 x 73 | 3496 | 46 x 76 | 3565 | 31 x 115 | 3636 | 36 x 101 |
| 3432 | 33 x 104 | 3498 | 33 x 106 | 3567 | 41 x 87 | 3638 | 34 x 107 |
| 3432 | 39 x 88 | 3498 | 53 x 66 | 3569 | 43 x 83 | 3640 | 35 x 104 |
| 3432 | 44 x 78 | 3500 | 35 x 100 | 3570 | 30 x 119 | 3640 | 40 x 91 |
| 3432 | 52 x 66 | 3500 | 50 x 70 | 3570 | 34 x 105 | 3640 | 52 x 70 |
| 3434 | 34 x 101 | 3502 | 34 x 103 | 3570 | 35 x 102 | 3640 | 56 x 65 |
| 3440 | 40 x 86 | 3503 | 31 x 113 | 3570 | 42 x 85 | 3645 | 45 x 81 |
| 3440 | 43 x 80 | 3504 | 48 x 73 | 3570 | 51 x 70 | 3648 | 32 x 114 |
| 3441 | 31 x 111 | 3510 | 30 x 117 | 3572 | 38 x 94 | 3648 | 38 x 96 |
| 3441 | 37 x 93 | 3510 | 39 x 90 | 3572 | 47 x 76 | 3648 | 48 x 76 |
| 3444 | 41 x 84 | 3510 | 45 x 78 | 3575 | 55 x 65 | 3648 | 57 x 64 |
| 3444 | 42 x 82 | 3510 | 54 x 65 | 3577 | 49 x 73 | 3649 | 41 x 89 |
| 3445 | 53 x 65 | 3515 | 37 x 95 | 3584 | 32 x 112 | 3650 | 50 x 73 |
| 3450 | 30 x 115 | 3519 | 51 x 69 | 3584 | 56 x 64 | 3652 | 44 x 83 |
| 3450 | 46 x 75 | 3520 | 32 x 110 | 3588 | 39 x 92 | 3654 | 42 x 87 |
| 3450 | 50 x 69 | 3520 | 40 x 88 | 3588 | 46 x 78 | 3654 | 58 x 63 |
| 3451 | 29 x 119 | 3520 | 44 x 80 | 3588 | 52 x 69 | 3655 | 43 x 85 |
| 3456 | 32 x 108 | 3520 | 55 x 64 | 3589 | 37 x 97 | 3657 | 53 x 69 |
| 3456 | 36 x 96 | 3525 | 47 x 75 | 3591 | 57 x 63 | 3658 | 31 x 118 |
| 3456 | 48 x 72 | 3526 | 41 x 86 | 3596 | 31 x 116 | 3658 | 59 x 62 |
| 3456 | 54 x 64 | 3526 | 43 x 82 | 3596 | 58 x 62 | 3660 | 60 x 61 |
| 3458 | 38 x 91 | 3528 | 36 x 98 | 3597 | 33 x 109 | 3663 | 33 x 111 |
| 3465 | 33 x 105 | 3528 | 42 x 84 | 3599 | 59 x 61 | 3663 | 37 x 99 |
| 3465 | 35 x 99 | 3528 | 49 x 72 | 3600 | 30 x 120 | 3666 | 39 x 94 |
| 3465 | 45 x 77 | 3528 | 56 x 63 | 3600 | 36 x 100 | 3666 | 47 x 78 |
| 3465 | 55 x 63 | 3531 | 33 x 107 | 3600 | 40 x 90 | 3672 | 34 x 108 |
| 3468 | 34 x 102 | 3534 | 31 x 114 | 3600 | 45 x 80 | 3672 | 36 x 102 |
| 3468 | 51 x 68 | 3534 | 38 x 93 | 3600 | 48 x 75 | 3672 | 51 x 72 |
| 3471 | 39 x 89 | 3534 | 57 x 62 | 3600 | 50 x 72 | 3672 | 54 x 68 |
| 3472 | 31 x 112 | 3535 | 35 x 101 | 3600 | 60 x 60 | 3675 | 35 x 105 |
| 3472 | 56 x 62 | 3536 | 34 x 104 | 3604 | 34 x 106 | 3675 | 49 x 75 |
| 3476 | 44 x 79 | 3536 | 52 x 68 | 3604 | 53 x 68 | 3680 | 32 x 115 |
| 3477 | 57 x 61 | 3538 | 58 x 61 | 3605 | 35 x 103 | 3680 | 40 x 92 |
| 3478 | 37 x 94 | 3540 | 30 x 118 | 3608 | 41 x 88 | 3680 | 46 x 80 |
| 3478 | 47 x 74 | 3540 | 59 x 60 | 3608 | 44 x 82 | 3685 | 55 x 67 |
| 3479 | 49 x 71 | 3542 | 46 x 77 | 3610 | 38 x 95 | 3686 | 38 x 97 |
| 3480 | 29 x 120 | 3549 | 39 x 91 | 3612 | 42 x 86 | 3689 | 31 x 119 |
| 3480 | 30 x 116 | 3550 | 50 x 71 | 3612 | 43 x 84 | 3690 | 41 x 90 |
| 3480 | 40 x 87 | 3551 | 53 x 67 | 3616 | 32 x 113 | 3690 | 45 x 82 |
| 3480 | 58 x 60 | 3552 | 32 x 111 | 3618 | 54 x 67 | 3692 | 52 x 71 |
| | | | | 3619 | 47 x 77 | | |

FIGURE 3.8. Two-Stage Gear Combinations

CHAPTER 4

# Inventions and Applications

John von Neumann famously joked, "Anyone who considers arithmetical methods of producing random digits is, of course, in a state of sin." [**233**]. The Farey sequence offers redemption.

The Farey sequence was regarded as being sufficiently **regular** by the table makers of the previous chapter to permit linear interpolation. The inventors and application builders of the current chapter regard it as being sufficiently **irregular** to be out of synchronization with any regularity. As we shall see, it is exactly this straddling of the border between regularity and irregularity that links the Farey sequence to the Riemann hypothesis.

Table 4.1 lists some patents and patent applications that refer to the Farey fractions and the mediant function.

TABLE 4.1. Patents and Patent Applications Using the Mediant

| Number | Title |
| --- | --- |
| US 4,542,336 | Method and apparatus for sampling broad band spectra in fuel quantity measurement systems [201] |
| US 5,253,192 | Signal processing apparatus and method for iteratively determining arithmetic Fourier transform [230] |
| US 5,708,432 | Coherent sampling digitizer system [208] |
| US 5,964,709 | Portable ultrasound imaging system [39] |
| US 6,477,553 | Measurement scale for non-uniform data sampling in N dimensions [63] |
| US 6,719,433 | Lighting system incorporating programmable video feedback lighting devices and camera image rotation [14] |
| US 6,840,439 | Fraction exploration device [10] |
| US 7,158,569 | Methods of digital filtering and multi-dimensional data compression using the Farey quadrature and arithmetic, fan, and modular wavelets [193] |
| US 2006/0159259 | Encryption and signature schemes using message mappings to reduce the message size [90] |
| US 2007/0115734 | Method of operating an integrated circuit tester employing a float-to-ratio conversion with denominator limiting [214] |
| US 2007/0208693 | System and method of efficiently representing and searching directed acyclic graph structures in databases [37] |
| US 2008/0065709 | Fast correctly-rounding floating-point conversion [122] |
| US 2009/0049299 | Data Integrity and non-repudiation system [142] |
| US 2009/0066719 | Image dithering based on Farey fractions [44] |
| US 2009/0136154 | System and method for image sensing and processing [19] |

## 4.1. Sampling Algorithm

United Status patent number $5, 708, 432$ issued on January 13, 1998, is titled "Coherent Sampling Digitizer System." The inventors are David Reynolds and Roman Slizynski. The assignee is Credence Systems Corporation, of Fremont, California. The patent addresses the problem of sampling of a high-frequency periodic signal at a desired granularity using equipment that isn't up to the task. To be more specific, let $F_t$ be the frequency of the signal that is to be sampled and suppose that one wishes to acquire on the order of $N$ samples per period for the purposes of a satisfactory digitization of the high-frequency signal. Not being up to the task means that the sampling frequency of the equipment, $F_s$, can't be set to $NF_t$.

The tried and true solution to this problem is to sample $M$ periods of the high-frequency signal at a realizable sampling of $F_s$ where

$$F_t/F_s = M/N$$

and

$$gcd(M, N) = 1.$$

For a number of reasons all addressed in the patent, the reality of testing equipment and situations is that there are constraints on the values to which $M$ and $F_s$ can be set. As a result the problem becomes finding a feasible value of $F_s$ and an acceptable value of $M$ so that $F_t/F_s$ is very close to $M/N$.

A feasible value of $F_s$ is one that is an integral division $K$ of a testing equipment master clock frequency, $F_{clk}$; i.e. $F_s = F_{clk}/K$ . With this notation, the problem addressed by the patent is as follows:

Find integers $N$, $M$, $K$ and $J$ subject to

$$N_{min} < N < N_{max}$$
$$M_{min} < N < M_{max}$$
$$K_{min} < K < K_{max}$$
$$P = M/J \text{ an integer}$$
$$Q = N(K/J) \text{ an integer}$$

so that

$$gcd(M, N) = 1$$

$$gcd(P, Q) = 1$$

and

$$P/Q = F_t/F_{clk}$$

The patented algorithm begins by computing the order of the Farey sequence, $F_m$ to be searched. This is maximum allowable value of Q, namely

$$m = Q_{max} = N_{max}K_{max}$$

The algorithm picks a feasible starting value in $F_m$ and shows how to move from one element in $F_m$ to another to converge to $F_t/F_{clk}$.

In the process of describing their digitization method the inventors were not unaware of the history of the Farey sequence and, in fact, deserve high praise for trying to get the history right. In column 6 they write the following:

> In 1802 the mathematician Haros suggested using a Farey series as an aid to rational approximation of an irrational number. For example we can approximate a number such as pi (3.14156...), by simply scanning down a Farey series computing the decimal equivalent of each term until we find the fraction ($1/7 = 0.142857...$) most closely approximating the fractional portion of pi, and then add the integer portion (3) of pi to the selected fraction to obtain the approximation.

It's almost as if Haros' paper were being cited as prior art. Sadly, just a little further on, in column 6, their grade of A has to be knocked down to an C. Describing how their algorithm uses the Farey series they say the following:

> The algorithm makes use of a relationship (discovered by Farey) between three successive terms $(P_L/Q_L P_M/Q_M, P_R/P_R)$ (sic) of a Farey series of order $Q_M$:
>
> $$P_M/Q_M = (P_L + P_R)/(Q_L + Q_R) = P_M/Q_M$$

Even if we can attribute the mangled mathematics to a patent attorney, after coming so close we must hold the inventors responsible for the misreported history.

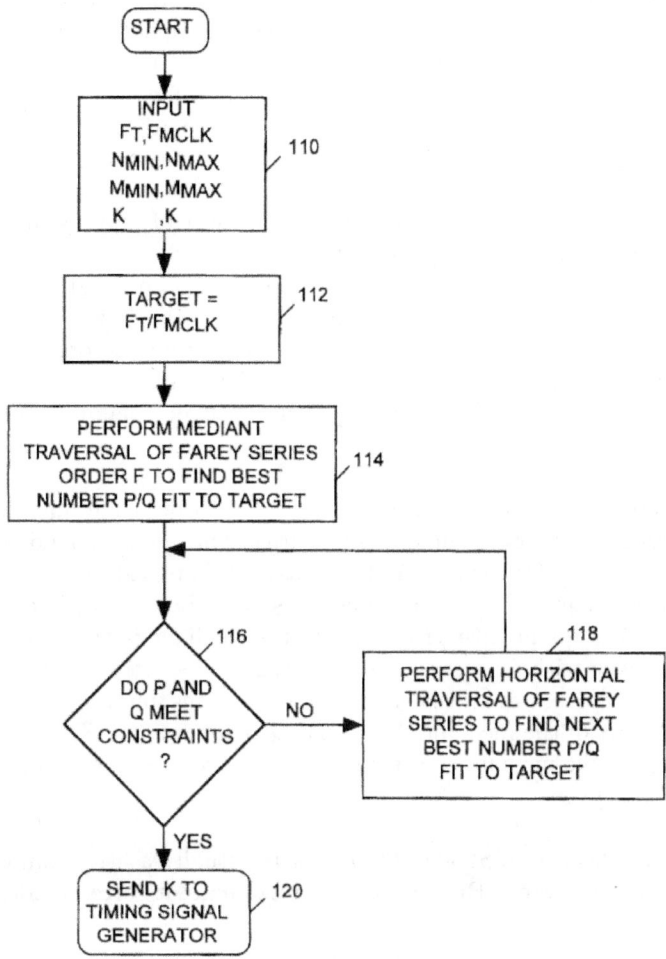

FIGURE 4.1. Figure 9 from US 5,708,432

## 4.2. Dithering Algorithm

United Status patent application number 2009/0066719A1 filed on September 7, 2007, is titled "Image Dithering Based on Farey Fractions." The inventor is Michael Combes and the assignee is Spatial Photonics, Inc., of Sunnyvale, California. The problem addressed by this patent application is the mapping of a picture from a high resolution display device to a low resolution device, for example from a desktop computer screen to a mobile telephone screen. The concern is not making the image smaller but rather mapping a palette $M$ color values onto a pallet of $N$ of color values where $N$ is much less than $M$.

Assuming the color values are put on some sort of linear scale – imagine black-and-white images with white at 0 and black at the top if that makes it easier – it's easy to posit a strictly increasing function from $[0, M]$ onto $[0, N]$ that does the job mathematically. The low-resolution image resulting from this straight-forward approach is often perceived by human observers to have edges where the high-resolution image didn't. These edges are called contours and dithering is the process of fuzzing them over and making them visually disappear.

The secret to a good dithering algorithm is to do something simple and repetitive without leaving behind any patterns. This almost sounds like a contradiction in terms. If it's simple and repetitive and applied to a pattern then it must create a pattern. The pattern may be simple, but it's a pattern nonetheless. One of the qualitative properties of the mediant is that its results are irregular, at least with respect to the human sensory system's built-in notion of regularity. Of course, we saw this in the Farey fractions, but it also seems to be a general property of the mediant. What patent application $US2009/0066719A1$ does is to apply the mediant to breaking up perceived regularity, namely edges, in graphics images.

Patent application $US2009/0066719A1$ is actually a little confused about Farey fractions and Stern-Brocot trees. For example, it correctly identifies the Farey sequence of order 5, $F_5$ as

$$F_5 = \{\frac{0}{1}, \frac{1}{5}, \frac{1}{4}, \frac{2}{5}, \frac{1}{2}, \frac{3}{5}, \frac{2}{3}, \frac{3}{4}, \frac{1}{1}\}$$

but then says

> For example, the term in $F_5$ that appears between $\frac{1}{3}$ (of $F_3$) and $\frac{1}{4}$ (of $F_3$) is $(1+1)/(3+4) = \frac{2}{7}$, as shown in FIG. 5.

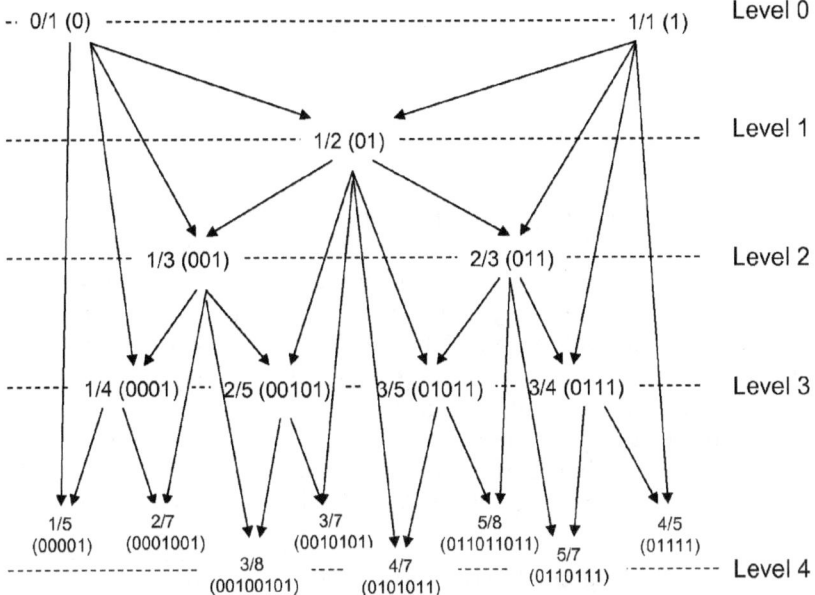

FIGURE 4.2. Figure 5 from US 2009/0066719 A1

whereas $\frac{2}{7}$ isn't in $F_5$ at all.

What is interesting about Figure 5 is not its tree structure but the binary version of the mediant that it defines. In particular, the patent application defines the binary representation of $\frac{a+c}{b+d}$ as being the binary representation of $\frac{a}{b}$ concatenated with the binary representation of $\frac{c}{d}$. Thus, to continue the example, the binary representation of the mediant of $\frac{2}{7}$, the (fractional) mediant of $\frac{1}{4}$ and $\frac{1}{3}$, is the binary representation of $\frac{1}{4}$, (0001) followed by the binary representation of $\frac{1}{3}$, (001), namely (0001001).

The sequence of 0's and 1's resulting from the binary representation of a particular sequence of Farey fractions is the dithering that is added back into the low-resolution image to smooth the edges. We've glossed over a number of details of the complete dithering algorithm. The reader is referred to the patent application itself for these details if they are of interest.

## 4.3. Decimal-to-Fraction Conversion

United Status patent application number $US2007/0115734A1$ filed on November 22, 2005, is titled "Method of Operating an Integrated Circuit Tester Employing a Float-to-Ratio Conversion with Denominator Limiting." The assignee is the same as the assignee of the above digitization patent, Credence Systems Corporation, of Fremont, California. The application sets out to patent the conversion of a number in decimal representation to the closest vulgar fraction with a denominator less that a given value; that is to patent exactly what Haros did 200 years before. Not surprisingly this time out Haros is not cited as prior art because this time Haros would represent disqualifying prior art. There are 3 independent claims and 4 dependent claims. Here's dependent claim 2:

> Claim 2. A method according to claim 1, wherein step (d) comprises
>
> comparing the result of a division to two consecutive members of a Farey series and determining whether either member is an acceptably close approximation of said result and, if not,
> calculating the mediate function of said two consecutive members and determining whether said mediant fraction is an acceptably close approximation of said result

Except for the explicit use of the Farey series, one might even cite Chuquet's règle des nombres moyens as prior art.

## 4.4. Analog-to-Digital Conversion

Gousuke Izawa, Toshimichi Saito and Hiroyuki Torikai published a paper entitled "A dependent switched capacitor A/D converter for Farey series approximation" in 2000 [141] that includes a control diagram, Figure 4.4, and circuit diagram, Figure 4.4, for an analog-to-digital converter based on Farey fractions. The authors begin by describing the basic $\Sigma - \Delta$ A/D converter

$$x_{n+1} = x_n - Q(x_n) + u$$

$$y_n = Q(x_n) = \begin{cases} 0 & x_n \leq 0 \\ 1 & x_n > 0 \end{cases}$$

$$\hat{u}_n = b^{-1} \sum_{i=0}^{b-1} y_{n+i}$$

The authors note that this converter yields a base $b$ digitization of the analog input $u$.

They go on to observe that if one is able to represent fractions with denominators up to and including $b$ on the output, it makes sense to use fractions with all of these denominators rather than limit oneself to fractions with denominators equal to $b$. To this end they posit the following digitization scheme based on a window $W_b = [-1/b, 1/b]$:

$$x_n = Q(x_n) = \begin{cases} x_n - Q(x_n) + u & x_n \notin W_b \\ u & x_n \in W_b \end{cases}$$

$$y_n = Q(x_n - 1/b)$$

$$\hat{u}_n = l^{-1} \sum_{i=0}^{l-1} y_{n+i}$$

The converter resets each time $x_n$ enters the window $W_b$ and $l$ counts the number of state transitions since the previous entry.

Figure 4.4 compares the base $b$ and Farey digitization of a linear input. Figure 4.4 compares the base $b$ and Farey digitization of a sine wave input. Figure 4.5 is the circuit diagram given in the Izawa paper that does Farey digitization.

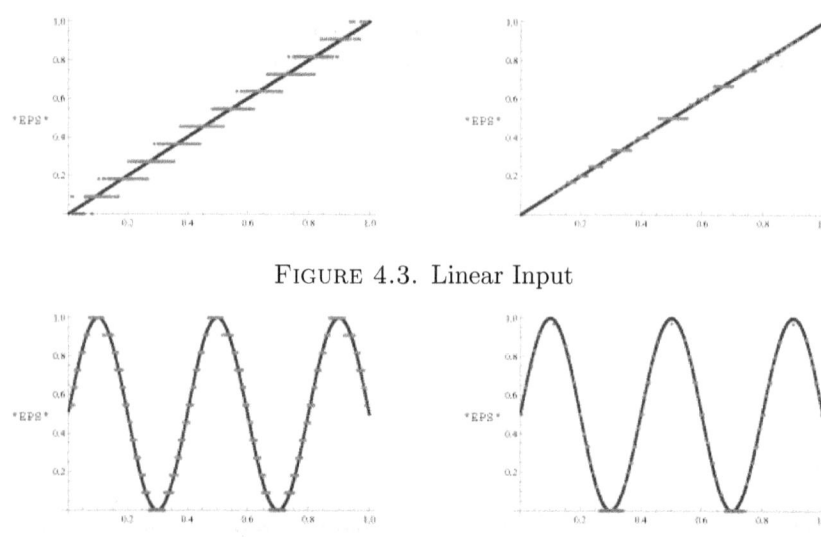

FIGURE 4.3. Linear Input

FIGURE 4.4. Sine Wave Input

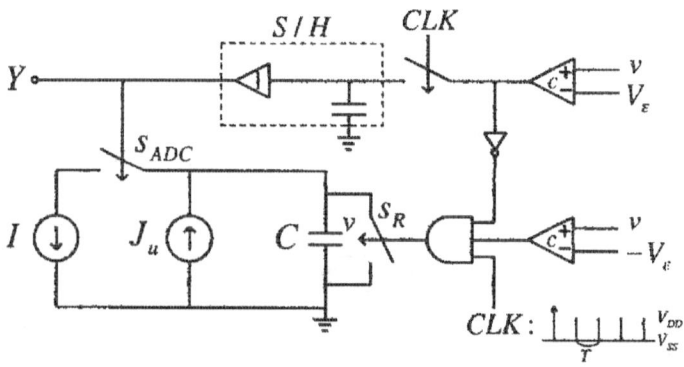

FIGURE 4.5. Circuit Diagram for Farey Digitization

## 4.5. Slash Arithmetic and Mediant Rounding

At the dawn of computing with digital electronics there were debates as to how numeric values should be represented as bits. Leading contenders included binary-coded-decimal (BCD), 2s-complement binary and continued fractions; roughly bankers versus electrical engineers versus mathematicians. Peter Henrici made the case for the mathematicians in 1956:

> Modern digital high-speed computation is unthinkable without the possibility of representing numbers by decimals. At the same time, the inherent error introduced by the representation of all numbers by a terminating decimal (or binary) fraction is a permanent source of some highly irritating features of digital computation. Numerical instability, loss of accuracy arising by subtraction of two nearly equal numbers, and round-off errors in general are a consequence of this ever-present inaccuracy. Little attention seems to have been paid to the simple idea that perfect accuracy is possible on a computing machine if the computations are performed in the field of rational numbers, that is, in the realm or fractions with integral numerator and denominator.[**135**]

Henrici was a guest worker at the U.S. National Bureau of Standards at the time. In the paper Henrici described a subroutine (a *method* in current parlance) for doing computations using rational numbers on the National Bureau of Standards Eastern Automatic Computer (SEAC). In the end the electrical engineers won. After all, they were the people building the machines. Nevertheless, like electric cars and the picturephone, the idea of computing with rational numbers keeps coming back, the argument being that since we now have all the gates we could ever desire, perhaps we could devote some attention to precision.

However, Haros and his co-workers at the Bureau du Cadastre have been successful far beyond their original charter and the world now thinks and works in the decimal system rather than fractions, vulgar or otherwise. As a result, the albatross around the neck of computing with rational numbers is the need to convert the decimal numbers that are now used universally in the street into and out of a rational representation. The nanosecond cost of back-and-forth translation per additional bit of precision realized has yet to prove economically attractive underneath Google Chrome, Microsoft Word or even Adobe Photoshop.

The second volume, *Seminumerical Algorithms*, of Knuth's monumental work, *The Art of Computer Programming* [**148**], is both a deep analysis and detailed history of all the ways one might represent numbers in computers including, of course, rational numbers. Many researchers have explored the pros and cons of computing with rational numbers but perhaps none more diligently or more persistently than David Matula and Peter Kornerup ([**173**], [**149**],[**174**]). Matula has over 15 patents in the field.    No matter how one represents numbers in a computer, one must step up to the issue of rounding. Even with today's gigabyte memory systems, sooner or later a calculation will try to create a number that doesn't fit and the arithmetic unit will return a result that is close but not exact. The properties of the rounding algorithm are what are at the core of any proposal for computer arithmetic.

When the arithmetic unit or the rational arithmetic software produces a computational result that cannot be represented in the memory allocated to each rational value, it uses the rounding algorithm to pick a number that can be represented as the (approximate) result. Since in this case the unrepresentable exact value will be strictly between two values that can be represented, the rounding algorithm only has to pick one of these with provision for a little extra bit of thinking if the exact result is exactly half way in-between.

It would seem hard to imagine a situation in which the best strategy wouldn't be rounding to the representable value that is closest to the non-representable value, but in the case of computing with rational numbers there are two arguments for setting the dividing line at the median of the two representable numbers rather than at their mean. Both arguments are based on valuing simple as well as accurate results (where $\frac{a}{b}$ is taken to be simpler than $\frac{c}{d}$ if $b < d$ or if $b = d$ then $a < c$) and the observation that median rounding produces simpler results than mean rounding.

The first argument is by that by rounding to a simpler result one will perform fewer roundings over the course of computation and therefore produce a more accurate final result. The second argument is that median rounding is more likely to find a simple exact final result if, in fact, the exact value is simple and it is worth tolerating a little extra error along the way to achieve this outcome.

The mediant rounding algorithm is as follows:

If $\frac{x}{y}$ is the non-representable exact result and $\frac{a}{b}$ and $\frac{c}{d}$ are the adjacent representable numbers, then round to $\frac{a}{b}$, if

$$\frac{a}{b} < \frac{x}{y} < \frac{a+b}{c+d} < \frac{c}{d},$$

and round to $\frac{c}{d}$

$$\frac{a}{b} < \frac{a+b}{c+d} < \frac{x}{y} < \frac{c}{d}.$$

If $\frac{x}{y}$ is equal to the mediant, then round to simpler of the two adjacent representable numbers.

While a number of generally agreed upon metrics exist for error – the absolute value and square of difference to just name two – it's a little harder to convert the above notion of simplicity to a single figure. Since a difference in the denominators trumps difference in numerators, to get a rough feel for the trade-off between error and simplicity we'll use the following to measure how much simpler $\frac{a}{b}$ is compared to $\frac{c}{d}$

$$Simpler(\frac{a}{b}, \frac{c}{d}) = \begin{cases} d - b & \text{if } b \neq d, \\ \frac{c-a}{2} & \text{if } b = d. \end{cases}$$

As an example of this entire process, suppose we implement fixed-slash arithmetic on an 8-bit microcontroller and allocated a byte each to the numerator and denominator of any value. A representable rational number $\frac{a}{b}$ would be such that $-128 < a < 128$ and $0 \leq b < 256$. Figure 4.6 is a log-log plot of the amount of simplicity yielded by mediant rounding against the difference between the mediant and mean rounded values for the 462 times that these two values differ in rounding the $48,610$ fractions less than 1 with the denominator of the $5,000^{th}$ prime.

Figure 4.7 is a plot of average (over the fractions where the mean rounding differed from the mediant rounding) error and simplicity for prime $p_i$, $i = 100(10)1000$. Roughly speaking, we are buying a one percent reduction in the denominator for a five-fold increase in the error.

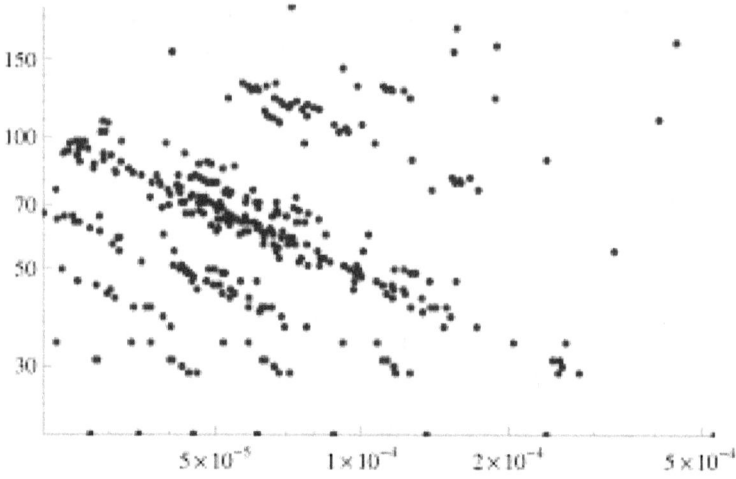

FIGURE 4.6. Error vs. Simplicity I

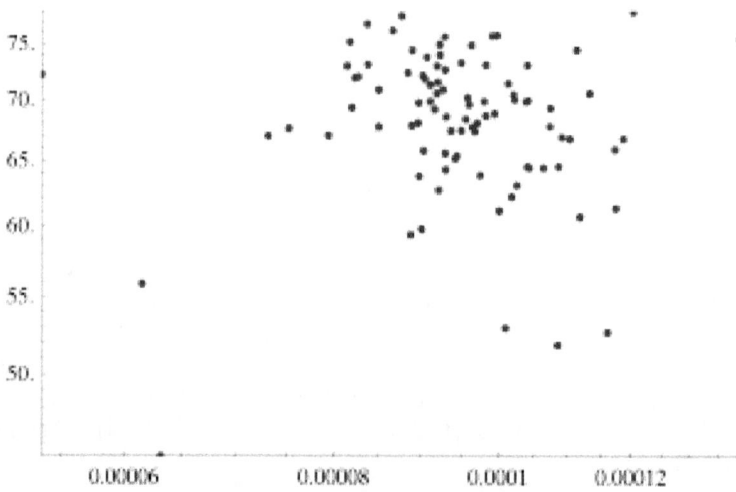

FIGURE 4.7. Error vs. Simplicity II

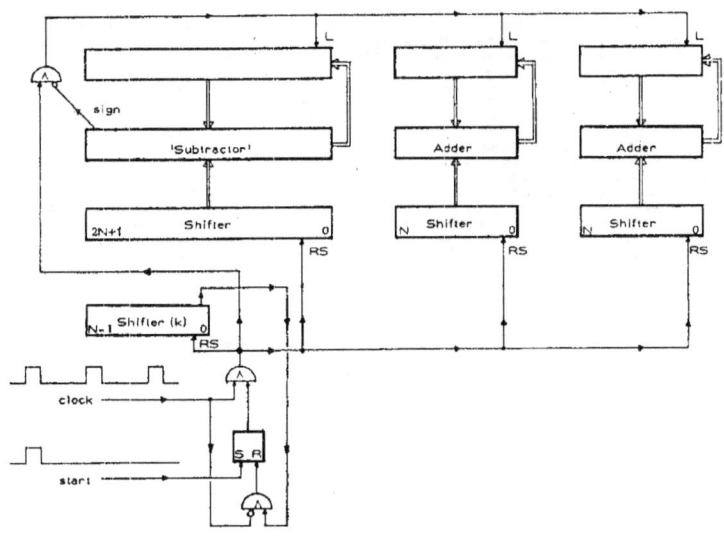

FIGURE 4.8. Mediant Rounding in Hardware

The arguments favoring mediant rounding are largely based upon experimental evidence, and even if they are appropriate for a particular problem
context, their promised value needs to be assessed in terms of the actual needs
for a solution. Scott [216] describes a fast version of the mediant rounding
algorithm that is incorporated in the MIRACL rational arithmetic software
package. Figure 4.8 is a sketch of a circuit for performing mediant rounding
that appears in [149]. Little is required beyond a couple of adders, a subtractor
and some barrel shifters.

## 4.6. Patterns for Weaving

Unlike contemplating a display of the primes, where it is hard to discern
clear patterns, it seems no matter how you display the Farey encoding of the
primes, patterns just jump out at you. For example, Figure 4.9 is a raw plot of
the Farey fractions against the order of the sequence in which they are found.
Coming up with equations that capture these patterns so that they can be

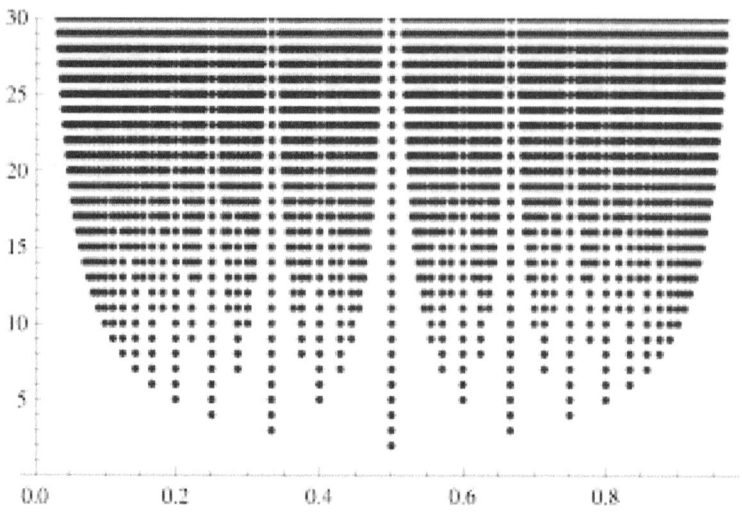

FIGURE 4.9. Farey Sequences, $m = 2(1)30$

analyzed is another story. However, patterns are useful even when they can't be packaged up in a tidy closed equation.

At Bell Laboratories in the mid 60's Ralph Griswold, along with David Farber and Ivan Polonsky, invented one of the many now-forgotten programming languages, Snobol. Snobol (String Oriented Symbolic Language) was – and still is for that matter since there is an implementation of SNOBOL in Python – a programming language specifically designed for describing, finding and manipulating patterns in strings of characters. Griswold left Bell Laboratories in 1971 to become the first professor of computer science at the University of Arizona, eventually heading the department. When he retired, his interest in patterns shifted from one-dimensional strings to two-dimensional weaving patterns. As a quick tour of any weaving archive will quickly convince you, weavers create and consume patterns at a voracious rate. Griswold was a prodigious contributor to the Digital Archive on Weaving, Lace and Related Topics. In addition to 147 patterns he produced two massive collections of weaving patterns. The "Historic Weaving Archive" is a 12-volume set and the other, "Historic Lace Archive", is an 8-volume collection.

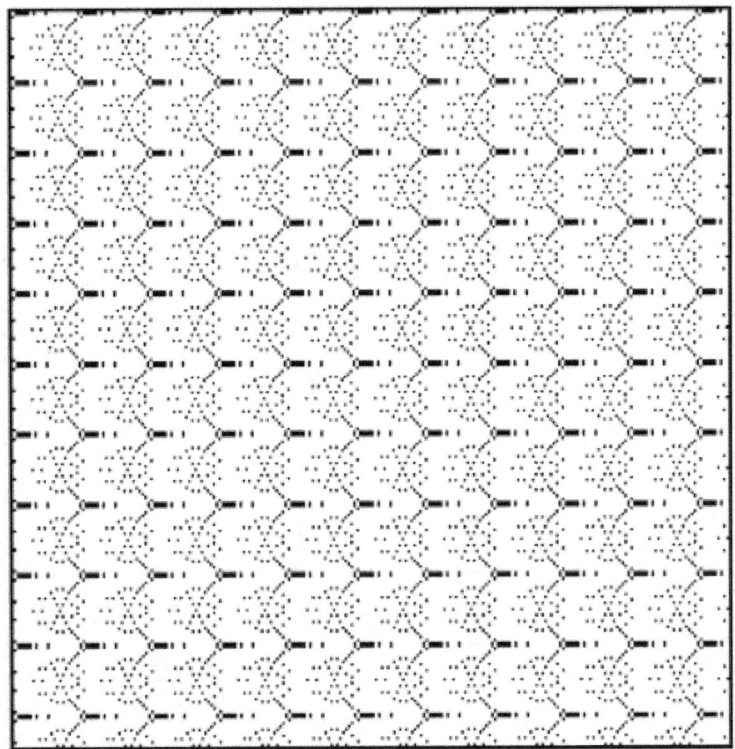

FIGURE 4.10.   Design using Numerators and Denominators of $F_8$

Griswold died in 2006 but, as of this writing, the University of Arizona computer science department still maintains Griswold's On-Line Digital Archive of Documents on Weaving and Related Topics. In this archive is a paper Griswold wrote in 2002 entitled "Designing with Farey Fractions". In the paper he describes a number of uses of the numerators and denominators of the Farey sequence to build weaving patterns. Figure 4.10 is a pattern built from the numerators and denominators of $F_8$. Figure 4.11 is a pattern that uses the denominators of $F_1$ through $F_8$.

Griswold finishes the paper by noting

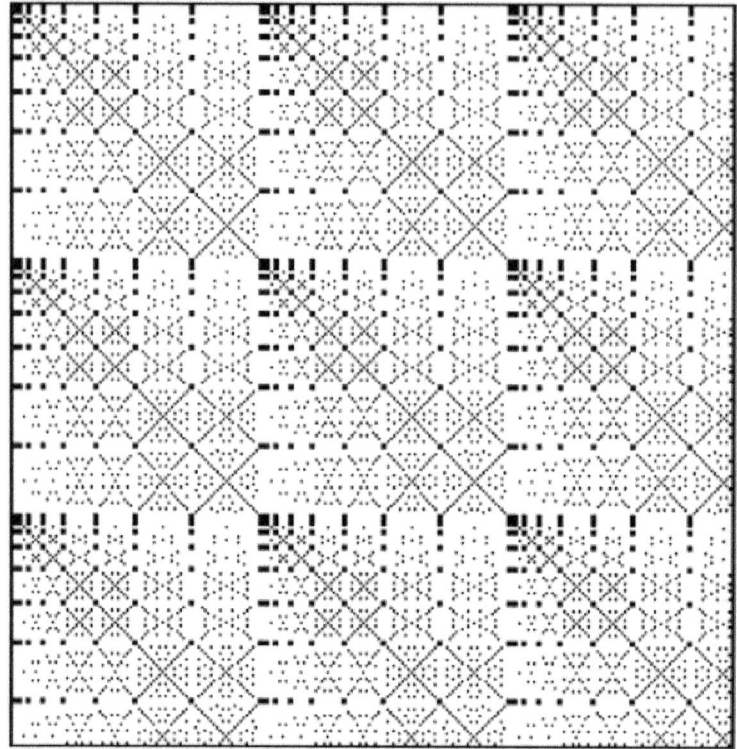

FIGURE 4.11.   Design using Denominators of $F_1$ through $F_8$

The wide range of possibilities touched on here does not begin
to exhaust the potential of Farey fraction design. Color specifi-
cation, for example, always is a design possibility for sequences.
[**120**]

The notion of using mathematics and, in particular, number theory to produce designs for textiles is not new. Anne-Marie Décaillot published a paper
in 2002 [**57**] that reviews the efforts of a group of nineteenth century mathematicians to demonstrate the relevance of mathematics to everyday affairs by
developing a wide range of number theoretic sequences for the design of fabrics

and mosaics. Like music, weaving is an ancient practice that has long been associated with mathematics.

### 4.7. Networks of Resistors

An electrical engineering puzzler of long standing asks for the number of effective resistances that can be constructed using $n$ given resistors. A simplified variant asks for the number of resistances that can be built using $n$ one-ohm resistors using only serial and parallel connections. Sameen Kahn in [143] observes that sets of effective resistances bear a useful relation to the Farey sequence that he harnesses to produce bounds on the answer to the puzzler.

Figure 4.12 shows the four possible configurations of three resistors and the effective resistance. First, Kahn shows that if $\frac{a}{b}$ is an effective resistance then so is $\frac{b}{a}$ so we can concentrate on effective resistances between 0 and 1.

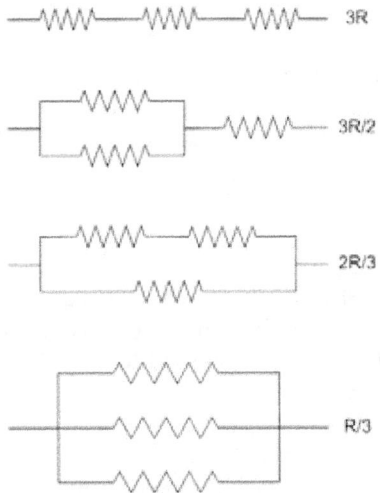

FIGURE 4.12.   Resistances with Three Resistors

Fewer effective resistances are possible if one has to use all $n$ resistors than if one is permitted to use no more than $n$ of the given resistors. Kahn considers both cases and as well as the case of bridge circuits. In the case that $n = 5$ and one must use all the resistors the effective resistances (the ones between 0 and 1) are

$$\left\{ \frac{1}{5}, \frac{2}{7}, \frac{3}{8}, \frac{3}{7}, \frac{1}{2}, \frac{4}{7}, \frac{5}{8}, \frac{5}{7}, \frac{4}{5}, \frac{5}{6}, \frac{6}{7} \right\}$$

If one can use at most 5 resistors then the effective resistances are

$$\left\{ \frac{1}{5}, \frac{1}{4}, \frac{2}{7}, \frac{1}{3}, \frac{3}{8}, \frac{2}{5}, \frac{3}{7}, \frac{1}{2}, \frac{4}{7}, \frac{3}{5}, \frac{5}{8}, \frac{2}{3}, \frac{5}{7}, \frac{3}{4}, \frac{4}{5}, \frac{5}{6}, \frac{6}{7}, 1, \right\}$$

The following code is adapted from the Mathematica code in Kahn's paper.

```
EffectiveResistances[resistors_, exactly_, countOnly_] :=
 Module[{configurations, resistances, S, SX},
  ClearAll[CirclePlus, CircleTimes];
  SetAttributes[{CirclePlus, CircleTimes}, {Flat, Orderless}];
  SeriesCircuit[a_, b_] := CirclePlus[a, b];
  ParallelCircuit[a_, b_] := CircleTimes[a, b];
  F[a_, b_] :=
   Flatten[Outer[SeriesCircuit, a, b] \[Union]
     Outer[ParallelCircuit, a, b], 2];
  S = {{R}, {CirclePlus[R, R], CircleTimes[R, R]}};
  Do[SX = F[S[[1]], S[[i - 1]]];
   Do[SX = Flatten[Union[SX, F[S[[k]], S[[i - k]]]], 2], {k, 2,
     i/2}];
   S = Union[S, {SX}], {i, 3, resistors}];
   configurations = Which[exactly , S[[resistors]],
       True, Union[Flatten[Table[S, {i, resistors}]]]];
  SetAttributes[{CirclePlus, CircleTimes},
              {NumericFunction, OneIdentity}];
  CirclePlus[a_, b_] := a + b;
  CircleTimes[a_, b_] := a*b/(a + b);
  resistances = Which[exactly , Union[configurations],
                 True, Union[configurations /. R -> 1]];
  If[countOnly,
   Return[{Length[configurations], Length[resistances]}],
   Return[{configurations /. R -> 1, resistances /. R -> 1}]
  ]]
```

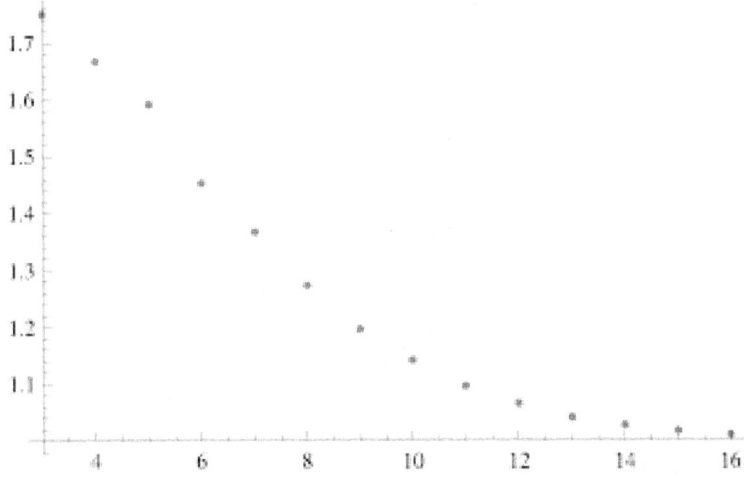

FIGURE 4.13. At-Most-n/Exactly-n

Let $A(n)$ be the number of effective resistances with exactly $n$ resistors. Kahn demonstrates the recurrance for $A(n)$ to establish a lower bound for $A(n)$ and computes the maximum denominator (or numerator) to establish an upper-bound, to wit

$$(1\sqrt{2})^n < A(n) < \Phi(\mathcal{F}_{n+1})$$

where $\mathcal{F}_k$ is the $k^{th}$ Fibonacci number. To this expression Kahn applies the well-known approximation for the totient and the closed form for the $k^{th}$ Fibonacci number to obtain

$$2.414^n < A(n) < 2.618^n$$

which he says agrees with computational results that yielded $2.55^n$ as the fit to $A(n)$ for $n = 2(1)16$.

Figure 4.13 plots the ratio of the number of effective resistances using at-most-$n$ and exactly-$n$ which suggests that this distinction vanishes with $n$.

The expression

$$lc(n) = 0.09 \exp 0.965n$$

TABLE 4.2. Farey-Based Estimate of Effective Resistances

| $n$ | $A(n)$ | $\hat{A}(n)$ | $A(n) - \hat{A}(n)$ |
|---|---|---|---|
| 4 | 4 | 3 | 1 |
| 5 | 11 | 8 | 3 |
| 6 | 26 | 21 | 5 |
| 7 | 65 | 57 | 8 |
| 8 | 168 | 152 | 16 |
| 9 | 434 | 404 | 30 |
| 10 | 1106 | 1070 | 36 |
| 11 | 2845 | 2835 | 10 |
| 12 | 7258 | 7509 | -251 |

is a reasonable fit to the length of the Farey sequence of order $\mathcal{F}_{n+1}$ and the set of effective resistances using $n$ resistors of $n$ and so

$$\hat{A}(n) = \frac{3}{\sqrt{5}\pi^2} \left( \varphi^{n+1} - \left( \frac{-1}{\varphi} \right)^{n+1} \right) - 0.09735 \exp 0.9528n$$

is another possible approximation for $A(n)$. Table 4.2 lists values of this approximation for $n = 4(1)12$.

# CHAPTER 5

# The Mediant and the Riemann Hypothesis

The Farey sequence encodes the prime numbers so it shouldn't surprise us that the modest sequence of vulgar fractions used by Charles Haros to build decimal conversion tables is intimately intertwined with the one of the longest outstanding problems in mathematics, the Riemann hypothesis. What may be a surprise is that the connection is just as simple and understandable as the series itself. The connection is exactly equivalent to the intuition about the series that Neville appealed to in building interpolation schemes for his Royal Society tables. This intuition was that the Farey sequence becomes a uniform division of the unit interval pretty quickly. Exactly how quickly is what binds the Farey sequence and the Riemann hypothesis.

In particular, if there is a constant value $M$ and a Farey sequence parameter $m_0$ such that for all values of the Farey parameter $m$ beyond $m_0$

$$m^{1-\epsilon} \sum_{f_i \in F_m} \left( f_i - \frac{i}{\varphi(n)} \right)^2 \leq M \tag{5.0.1}$$

then the Riemann hypothesis is true. This is a deceptively simple statement of an equivalent of the Riemann hypothesis. In case you didn't spot it, the deceptive bit is the $i$. It appears to be one variable but it is actually two. The subscript on $f$ is an order value and the numerator of the fraction is an arithmetic value. Equation 5.0.1 says that Farey sequence becomes regular almost as fast as $\frac{1}{n}$ goes to zero. Some of those who have answered the siren call of the Riemann hypothesis believe that the heart-of-hearts of the hypothesis, where doors will open to wholly new mathematical constructs, is in the 'almost'.

Figure 5.1 is a log plot of the $76,117$ terms in this sum for $m = 500$. Figure 5.2 is a plot of sum for $m = 100(20)500$ and $\frac{1}{n}$.

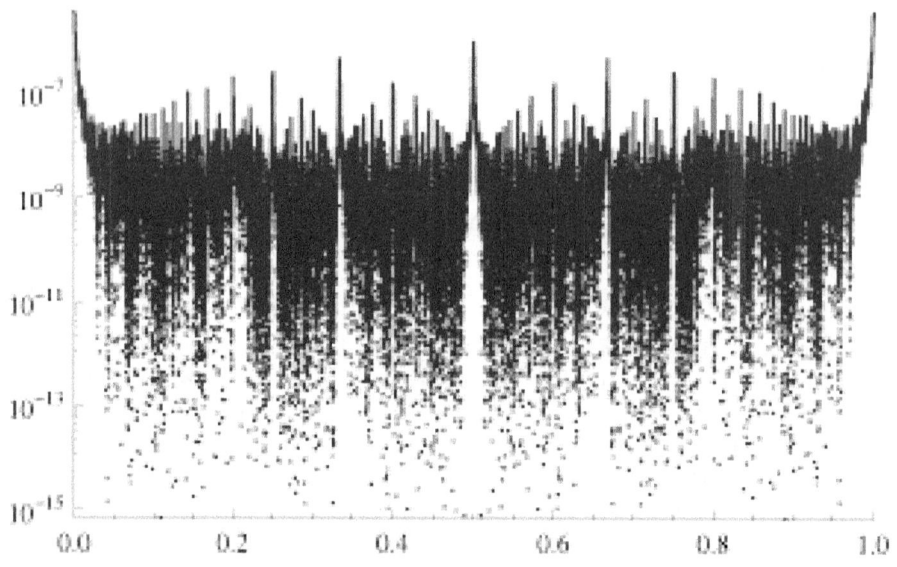

FIGURE 5.1. $ln(f_i - i/\Phi(n))^2, f_i \in F_{500}$

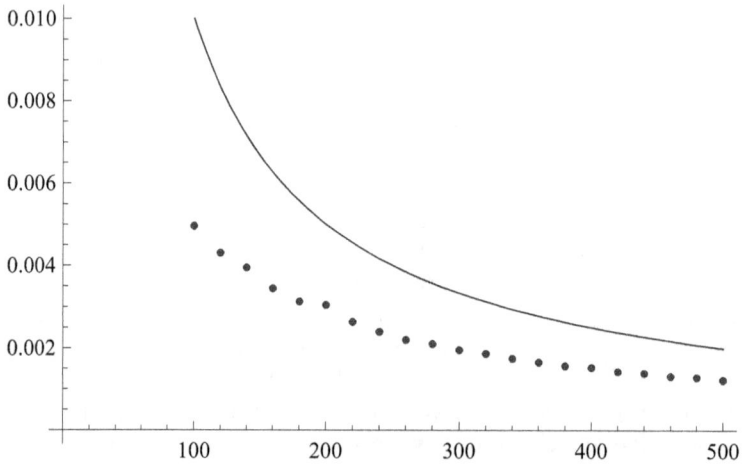

FIGURE 5.2. Sum of squared differences and $1/n$, $m = 100(20)500$

The rate at which the Farey sequence becomes regular is perhaps the easiest to grasp of all the expressions of the Riemann hypothesis. The proof of the Farey sequence connection can be understood by mathematics undergraduates and is only 600 words long. It was made by Jérôme Franel in a 4-page paper in the prestigious German mathematics journal, Göttinger Nachrichten, in 1924 [85]. That the relationship between a series of fractions so simple can be connected to a mathematical hypothesis so profound with such economy is the mark of a teacher of mathematics of the very highest order.

Another equivalence between a condition of the Farey sequence and the Riemann hypothesis, also accessible to the undergraduate, was published by Charles Pisot in 1960 [198]. Steckin summarized the state of attacks on the Riemann hypothesis by way of the Farey sequence up to 1997 [223]. Since that time Shigeru Kanemitsu and Masami Yoshimoto have published a series of papers exploring this connection much more deeply ([243], [244], [144], [245], [246])as has Akio Fujii ([86], [87]).

## 5.1. Jérôme Franel, Chair for Mathematics in the French Language

In a review of a book by Ferdinand Gonseth in the Spring-Summer 2006 (Volume XI, Issue 1) of the newsletter, International Society for the History of Philosophy of Science it is said that Gonseth was the "successor on Jerome's (sic) Franel's chair for Mathematics in French language at the ETH." The listing for Jérôme Franel at the Eidgenössiche Technische Hochschule Zürich Web page entitled "Liste aller Professoren" says of Franel: "1886-1929 Ordinarius für Mathematik in französischer Sprache. Direktor des Poly 1905-1909." A check with the current head of the mathematics department at ETH, Professor Dr. Giovanni Felder, confirmed that indeed until about 1968 there was an official chair for mathematics in the French language.

By all accounts Franel was first and foremost a teacher of mathematics. A shining moment in his life in the international mathematics community must have been when he was called upon to read Henri Poincaré's, opening lecture, "Sur les rapports de l'analyse pur et del la physique mathématique," at the first International Congress of Mathematicians held in Zürich in 1897. He was in his late 30's and had been at ETH only one year.

George Polya is quoted by Gerald Alexanderson in describing Franel as follows:

> He was an especially attractive kind of person and a very good teacher. He gave the introductory lectures on calculus in French for several decades. He had a real interest in mathematics, but he was more interested in French literature. Teaching occupied a good deal or his time but in French literature he had to read everything available. He had no time left to do mathematics. But when he retired he suddenly tackled two of the great problems: the Riemann hypothesis and 'Fermat's Last Theorem.' He asked his good friend Kollros and me one day to listen to his explanation of how he wished to prove the Riemann hypothesis. I listened and tried not to interrupt, but at one point I asked for an explanation. He stopped, was silent for a few minutes, then said. 'Yes, there is the error.' [1]

Franel published his "Les suites" paper between two other publications by other mathematicians that are of direct relevance. The one that appeared before Franel's provided the Riemann hypothesis equivalence that Franel harnessed. It was Littlewood's paper "Some consequences ..." [166]. The paper that appeared shortly after Franel's was Edmund Landau's "Bermerkungen zu der vorstehenden Abhandlung von Herrn Franel" [157]. Landau's paper was published in the same journal as Franel's and provided extra details about a function at the core of Franel's proof as well as an additional equivalence to the Riemann hypothesis. Because this second paper also concerned the rate of regularization of the Farey sequence, the linkage between the Farey sequence and the Riemann hypothesis is often credited jointly to Franel and Landau but Landau seems to grant full credit to Franel.

Franel's paper and Landau's paper follow directly below. The extensive first footnote in Landau's paper together with Littlewood's paper are included in the Appendix. The Landau footnote proves an identity used in Franel's paper that has become known as the two-dimensional Franel integral and found to be generally useful in analytic number theory.

The function $\mu$ referred to in the papers is the Möbius $\mu$ function:

$$\mu(n) = \begin{cases} 1 & \text{if } n \text{ is square-free with an even number of prime factors} \\ -1 & \text{if } n \text{ is square-free with an odd number of prime factors} \\ 0 & \text{if } n \text{ is not square-free} \end{cases}$$

## 5.2. "The Farey Series and the Prime Numbers Problem"

The Farey Series and the Prime Numbers Problem
by
J. **Franel** in Zürich
Presented by E. Landau at the November 21, 1924 meeting.

The hypothesis with respect to the imaginary roots of the function $\zeta(s)$ enunciated by Riemann in his famous treatise about prime numbers has never been either confirmed or invalidated. Its proof presents the greatest difficulties and appears to be reserved for the still distant future. In the meantime, we would propose in this note to demonstrate that the substance of the hypothesis in question is completely expressed by a property of the Farey series, where only rational numbers occur.

The Farey series of the $n^{th}$ order is constituted by the positive irreducible fractions whose denominator is at most equal to $n$.

Let

$$\varrho_1, \varrho_2, \varrho_3, \ldots$$

be such fractions arranged in increasing order,

$$A(n) = A = \varphi(1) + \varphi(2) + \ldots + \varphi(n)$$

the number of such fractions $\leq 1$, and more generally, $g(x, n)$, the number of terms in our series, is $\leq x$ , where $x$ designates any positive number.

The number of irreducible positive fractions $\leq x$ having the denominator $r$ can be expressed by the equation:

$$\sum \mu \left(\frac{r}{d}\right) [dx],$$

where the summation extends to the divisors of the number $r$. From this, we can conclude that

$$g(x,n) = \sum_{r=1}^{r=n} \sum_{d} \mu\left(\frac{r}{d}\right) [dx] = \sum_{a=1}^{a=n} [ax] M\left[\frac{n}{a}\right],$$

where

$$M(n) = \sum_{r=1}^{r=n} \mu(r).$$

Let us introduce the periodic function

$$f(x) = [x] - x + \frac{1}{2};$$

It will follow, if we observe that

$$A = \sum_{a} aM\left[\frac{n}{a}\right], \quad 1 = \sum_{a} M\left[\frac{n}{a}\right],$$

$$g(x,n) = \sum_{a} f(ax) M\left[\frac{n}{a}\right] - \frac{1}{2} + Ax,$$

or

$$g(x,n) = \delta(x,n) - \frac{1}{2} + Ax,$$

if we make

$$\delta(x,n) = \sum_{a} f(ax) M\left[\frac{n}{a}\right].$$

By $\varrho_0$, I mean the number 0. In the interval $\varrho_i \ldots \varrho_{i+1}$,

$$\delta(x,n) = i + \frac{1}{2} - Ax$$

or

$$\delta(x,n) = \delta_i - A(x - \varrho_i),$$

where, for the sake of brevity, one writes

$$\delta_i = i + \frac{1}{2} A \varrho_i,$$

thereby designating by $\delta_i$ the limit of $\delta(x,n)$ as $x$ tends towards $\varrho_i$ by decreasing values.

It is clear that we have

$$\delta_i + \delta_{A-i} = 1, \tag{5.2.1}$$

$$\delta_i + \delta_{A-i-1} = A(\varrho_{i+1} - \varrho_i), \tag{5.2.2}$$

from which we can specifically deduce that

$$\sum_{i=0}^{A-1} \delta_i = \frac{A}{2}.$$

Let us now consider the integral

$$I(n) = \int_0^1 \delta^2(x, n)dx.$$

We can easily ascertain [1]) that

$$\int_0^1 f(ax)f(bx)dx = \frac{(a,b)^2}{12a \cdot b}.$$

[1]) Using the progression of f (x) in a trigonometric series.

In this equation, $a$ and $b$ designate two positive whole numbers and $(a, b)$ their greatest common divisor.

This yields the result that

$$I(n) = \frac{1}{12} \sum_{b=1}^{b=n} \sum_{a=1}^{a=n} \frac{(a,b)^2}{a \cdot b} M\left[\frac{n}{a}\right] M\left[\frac{n}{b}\right]$$

Let us make

$$C(n) = \sum_{a=1}^{a=n} \frac{1}{a} M\left[\frac{n}{a}\right]$$

and

$$H(x) = \sum{}' \frac{1}{a \cdot b} M\left[\frac{n}{a}\right] M\left[\frac{n}{b}\right],$$

the summation in this last equation extending to the pairs of whole positive numbers $a$ and $b \leq n$, and primes between them.

We will have

$$C^2(n) = \sum_r \frac{1}{r^2} H\left[\frac{n}{r}\right],$$

from which

$$H(n) = \sum_r \frac{\mu(r)}{r^2} C^2\left(\left[\frac{n}{r}\right]\right),$$

and thus

$$I(n) = \frac{1}{12} \sum_r H\left[\frac{n}{r}\right],$$

$$I(n) = \frac{1}{12} \sum_r \sigma(r) C^2\left[\frac{n}{r}\right], \tag{5.2.3}$$

where

$$\sigma(n) = \sum_d \frac{\mu(d)}{d^2} = \prod_p \left(1 - \frac{1}{p^2}\right).$$

The summation extends to the divisors of $r$ and the product of the prime factors contained in $r$.

The quantities $\sigma(r)$ are thus the positive values $\leq 1$. As for $C(n)$, this represents the sum of the $n$ first coefficients in the development of the function $\frac{\zeta(s+1)}{\zeta(s)}$ in the Dirichlet series:

$$\frac{\zeta(s+1)}{\zeta(s)} = \sum_{n=1}^{\infty} \frac{c(n)}{n^2},$$

$$C(n) = \sum_{r=1}^{r=n} c(r).$$

On the other hand, we have

$$I(n) = \sum_{i=0}^{A-1} \int_{\varrho_i}^{\varrho_{i+1}} \delta^2(x,n)dx = \sum_i \int_{\varrho_i}^{\varrho_{i+1}} (\delta_i - A(x - \varrho_i))^2 dx,$$

from which, taking equation (2) into consideration,

$$I(n) = \sum_{i=0}^{A-1} \left(\frac{\delta_i^3 + \delta_{A-i-1}^3}{3A}\right) = \frac{2}{3A} \sum_i \delta_i^3.$$

In addition, because of equation (1), $\delta_i - \frac{1}{2}$ simply changes sign when we replace $i$ with $A - i$, such that

$$\sum_i \left(\delta_i - \frac{1}{2}\right)^3 = 0,$$

from which, since $\sum_i \delta_i = \frac{A}{2}$, it follows that

$$I(n) = \frac{1}{A}\sum_i \delta_i^2 - \frac{1}{6},$$

which we can express in the form

$$\sum_i \left(\delta_i - \frac{1}{2}\right)^2 = A\left(I(n) - \frac{1}{12}\right)$$

or

$$\sum_i \left(\frac{i}{A} - \varrho_i\right)^2 = \frac{1}{A}\left(I(n) - \frac{1}{12}\right)$$

Now, if the Riemann hypothesis is valid, $M(n)$, as demonstrated by M. Littlewood [2]), is of the order of $n^{\frac{1}{2}+\epsilon}$ (where $\epsilon$ represents a positive value that can be as small as one wishes); it is clearly the same for $C(n)$. As a consequence, $I(n)$, because of equation (3), is of the order $n^{1+\epsilon}$.

Reciprocally, if $I(n)$ is of the order $n^{1+\epsilon}$, then we will have, a fortiori,

$$C^2(n) = \mathrm{O}\left(n^{1+\epsilon}\right),$$

$$C^n) = \mathrm{O}\left(n^{\frac{1}{2}+\epsilon}\right).$$

The series

$$\frac{\zeta(s+1)}{\zeta(s)} = \sum_n \frac{c(n)}{n^2}$$

converges for $\Re(s) > \frac{1}{2}$ and Riemann's hypothesis is confirmed.

The renowned geometrist's assertion thus comes down to the following:

$$I(n) = \mathrm{O}\left(n^{1+\epsilon}\right)$$

or, which amounts to the same thing, to this:

$$\sum_i \left(\frac{i}{A} - \varrho_i\right)^2 = \mathrm{O}\left(\frac{1}{n^{1-\epsilon}}\right).$$

This is exactly what we set out to establish.

[2]) Proceedings of the Paris Academy of Sciences, vol. 154 (1912) p. 263-206.

### 5.3. A Synopsis of Franel's Proof

Franel uses the following Riemann equivalence to prove his own:

THEOREM 2. *The Riemann hypothesis is true if and only if the series*

$$\frac{\zeta(s+1)}{\zeta(s)} = \sum \frac{c(n)}{n^s}$$

*where*

$$c(n) = \sum_{d|n} \frac{d}{n} \mu(d) = \frac{1}{n} \prod_{p|n}(1-p)$$

*converges for* $\Re(s) > \frac{1}{2}$.

To do so he has to connect the uniformity measure

$$\sum_{f_i \in F_m} \left(f_i - \frac{i}{\varphi(n)}\right)^2$$

to

$$\sum \frac{c(n)}{n^s}$$

He does this by means of the function $I(n)$. Think $I(n)$ as an island in the middle of a river.

On one side he builds a bridge between the uniformity measure and $I(n)$:

$$\sum_{f_i \in F_m} \left(f_i - \frac{i}{\Phi(n)}\right)^2 = \frac{1}{\Phi(n)}\left(I(n) - \frac{1}{12}\right)$$

This is accomplished in the section of the paper starting with "On the other hand ..." using the $\delta_i$. A integral part of this of this bridge is the function $\delta_i$ a plot of which is given in Figure 5.3.

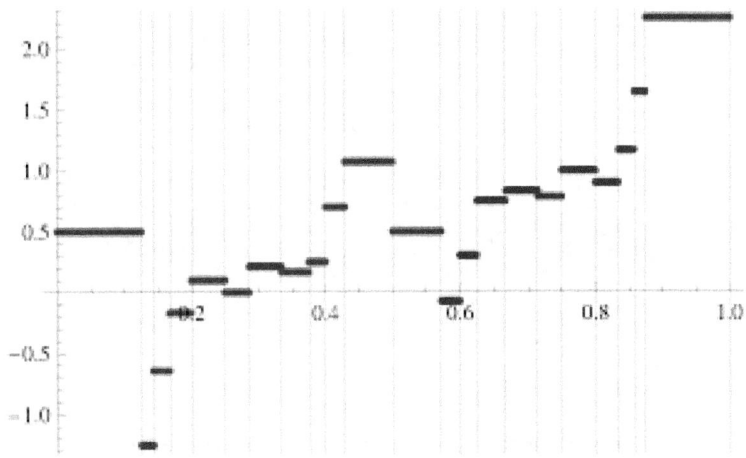

FIGURE 5.3. $\delta_i(x)$ for $r = 8$

On the other side he builds a bridge between $I(n)$ and the $c(n)$ using $C(n)$. After establishing that

$$I(n) = \frac{1}{12} \sum_r \sigma(r) C^2 \left[\frac{n}{r}\right]$$

he notes that we can ignore the $\sigma(r)$ terms and – one imagines a sparkle in his eye – that

$$C(n) = \sum_{i=1}^{n} c(n).$$

Figures 5.4 and 5.5 are plots of the parts of this second bridge, namely $c(n)$ and $C(n)$ respectively. Figure 5.6 is a plot of the island in the middle of the river, $I(n)$.

The second bridge gets him $I(n) = (O)(n^{1+\varepsilon})$ and $1/\Phi(n) = \mathcal{O}(\frac{1}{n^2})$ the first bridge gets him $O(1/n^{2-(1+\varepsilon)})$.

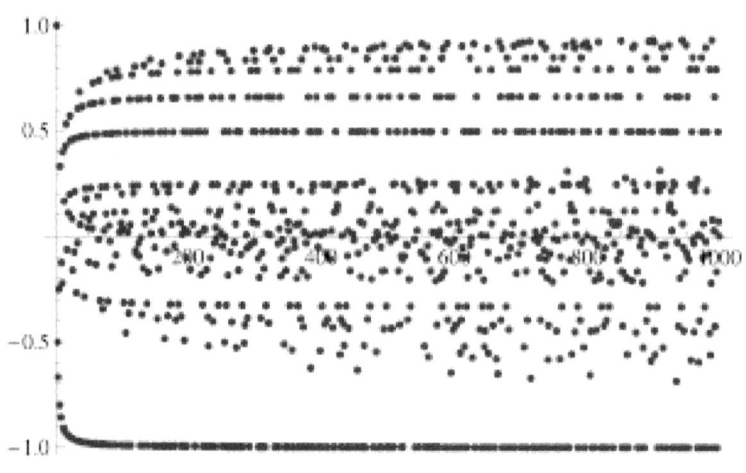

FIGURE 5.4. $c(n)$ for $n = 1(1)1000$

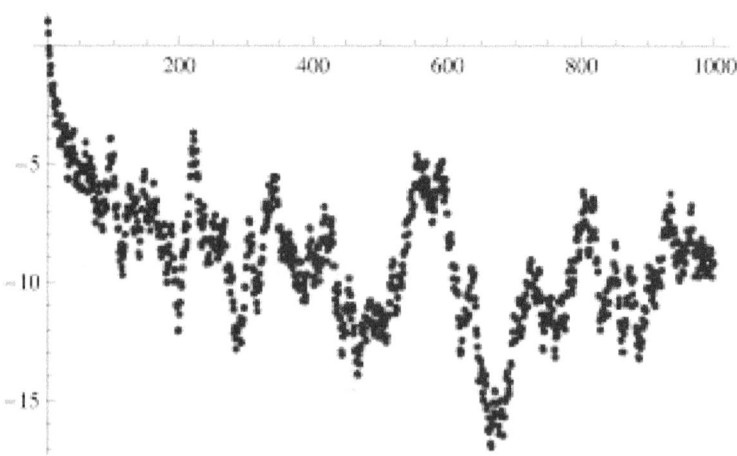

FIGURE 5.5. $C(n)$ for $n = 1(1)1000$

A much more complete and elegant recitation of the connection between the Farey sequence and the Riemann hypothesis is given in Section 12.2 of the book *Riemann's Zeta Function* by Edwards [65].

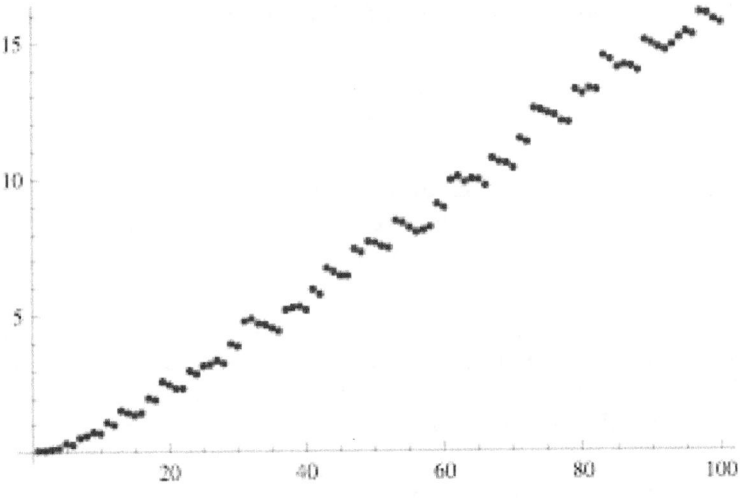

FIGURE 5.6. $I(n)$ for $n = 1(1)100$

In the author's opinion this is the one of the most elegant and stunning proofs in the entire history of the theory of numbers. It is not at all hard to understand why Franel sensed that the gold ring of proving the Riemann hypothesis was within his reach.

Landau's paper refined Franel's approach to prove a variant of Franel's result; namely that the Riemann hypothesis is true if and only if

$$\sum_{f_i \in F_m} |f_i - \frac{i}{\varphi(n)}| = \mathcal{O}(n^{1/2+\epsilon})$$

## 5.4. "Remarks Concerning the Earlier Paper by Mr. Franel"

Remarks concerning the earlier paper by
Mr. Franel
By
**Edmund Landau**
Presented at the November 21, 1924 meeting.
**Introduction**

It has been known for the past twelve years thanks to Mr. Littlewood that the Riemannian hypothesis regarding the zeroing of the zeta function is equivalent to a property of the Farey series. For Mr. Littlewood had shown that the Riemannian hypothesis is equivalent to

$$M(n) = \sum_{r=1}^{n} \mu(r) = O(n^{\frac{1}{2}+\varepsilon})$$

(for every $\varepsilon > 0$); on the other hand, it is generally known that $\mu(r)$ is the sum of the prime $r$-th roots of unity, thus

$$M(n) = \sum_{\nu=1}^{A} \exp 2\pi i \rho_\nu = \sum_{\nu=1}^{A} \cos 2\pi \rho_\nu,$$

where $\rho_\nu$ intersects with the positively ordered reduced series $\leq 1$ with denominators $\leq n$, and $A = A(n) = \sum_{r=1}^{n} \varphi(r)$ represents their number. Thus we knew that

$$\sum_{\nu=1}^{A} \exp 2\pi i \rho_\nu = O(n^{\frac{1}{2}+\varepsilon})$$

was equivalent to the Riemannian hypothesis.

Mr. Franel's great achievement is that he demonstrated that:

If we assume that

$$\rho_\nu - \frac{\nu}{A} = \eta_n = \eta_\nu(n)$$

then the Riemannian hypothesis is equivalent to the rational relationship

$$\sum_{\nu=1}^{A} \eta_\nu(n)^2 = O(n^{-1+\varepsilon})$$

between Farey series.

The principal part of Franel's proof consists of the presentation of the following identity (6), from which he easily concludes the rest on the basis of Littlewood's theorem : If we assume that

$$C(n) = \sum_{a=1}^{n} \frac{1}{a} M\left[\frac{n}{a}\right],$$

$$I = I(n) = \frac{1}{12} \sum_{r=1}^{n} C^2\left[\frac{n}{a}\right] \prod_{p|r}(1 - \frac{1}{p^2})$$

then it follows that

$$\sum_{\nu=1}^{A} \eta_\nu(n)^2 = \frac{1}{A}\left(I - \frac{1}{12}\right).$$

To provide some orientation, I should note at the same time that the trivial estimations

$$M(n) = O(n), \qquad \frac{1}{A(n)} = O\left(\frac{1}{n^2}\right)$$

from (4), (5) and (6) yield:

$$C(n) = \sum_{r=1}^{n} \frac{1}{a}\frac{n}{a} = O(n),$$

$$I = O\sum_{a=1}^{n} \frac{n^2}{r^2} = O(n^2),$$

$$\sum_{\nu=1}^{A} \eta_\nu^2 = O(1).$$

In this note, I would like to add three different points:

1) The Riemannian hypothesis is equivalent to the relationship

$$\sum_{\nu=1}^{A} |\eta_\nu| = O\left(n^{\frac{1}{2}}\right)$$

For this purpose, I have only to demonstrate in addition that:

I) (8) follows from (3),

II). (2) follows from (8).

For according to Mr. Franel, (3) follows from (2). 2). The relationship

$$\sum_{\nu=1}^{A} \exp 2\pi i \frac{\nu}{A} \eta_n u = O\left(n^{\frac{1}{2}+\varepsilon}\right)$$

(which obviously implying nothing more than (8), is equivalent to the Riemannian hypothesis, and thus, according to (1), implies nothing less than (8).

In addition, I only have to demonstrate that (2) follows from (9).

3). Each of the two relations

$$\sum_{\nu=1}^{A} \eta_\nu^2 = O\left(n^{-1+\frac{21\,\lg\lg\lg n}{\lg\lg n}}\right)$$

and

$$\sum_{\nu=1}^{A} |\eta_\nu| = O\left(n^{\frac{1}{2}+\frac{11\,\lg\lg\lg n}{\lg\lg n}}\right)$$

(which, obviously implying nothing less than (3) or (8)), is equivalent to the Riemannian hypothesis, and according to the Franelian theorem, imply nothing more than (3) or according to (1), nothing more than (8).

In addition, I only have to demonstrate that (10) and (11) follow from the Riemannian hypothesis.

I). According to Cauchy's inequality, and considering that $A \leqq \sum_{\nu=1}^{n} \nu \leqq n^2$, it follows that

$$\sum_{\nu=1}^{A} |\eta_\nu| \leqq \sqrt{\sum_{\nu=1}^{A} 1^2 \cdot \sum_{\nu=1}^{A} \eta_\nu^2} = \sqrt{A} \sqrt{\sum_{\nu=1}^{A} \eta_\nu^2} \leqq n \sqrt{\sum_{\nu=1}^{A} \eta_\nu^2}$$

$$= O\left(n^{1+\frac{-1+\varepsilon}{2}}\right) = O\left(N^{\frac{1}{2}+\varepsilon}\right)$$

thus (8). (The size of $\eta_\nu$ was not used here.)

II) For $n > 1$,

$$\sum_{\nu=1}^{A} \exp 2\pi i \frac{\nu}{A} = 1,$$

$$\sum_{\nu=1}^{A} \exp 2\pi i \rho^\nu = \sum_{\nu=1}^{A} \exp 2\pi i \frac{\nu}{A} \exp 2\pi i \eta_\nu = \sum_{\nu=1}^{A} \exp 2\pi i \frac{\nu}{A} \left(\exp 2\pi i \eta_\nu - 1\right),$$

$$\left| \exp 2\pi i \frac{\nu}{A} \exp 2\pi i \rho_\nu \right| \leqq \exp 2\pi i \frac{\nu}{A} |\exp 2\pi i \eta_n u - 1|$$

$$= 2 \exp 2\pi i \frac{\nu}{A} |\sin pi \eta_n u| \leqq 2\pi \exp 2\pi i \frac{\nu}{A} |\eta_\nu|.$$

Thus, (2) also follows from (8).

The amusing thing is that (3) also follows roundabout from (8) (by way of (2)); something that one of course cannot evaluate without reference to the size of the numbers For $-2\pi \leqq \alpha \leqq 2\pi$,

$$|\exp \alpha i - 1 - \alpha i| \leqq \sum_{m=2}^{\infty} \frac{|\alpha|^m}{m!} \leqq c\alpha^2.$$

Thus, from (13) and (7) it follows that

$$\left| \sum_{\nu=1}^{A} \exp 2\pi i \rho_n u - 2\pi i \sum_{\nu=1}^{A} \exp 2\pi i \frac{\nu}{A} \eta_\nu \right| \leqq 4\pi^2 c \sum_{\nu=1}^{A} \eta_\nu^2 = O(1).$$

In the event that (9) is true, then (2) is also true.

In a paper that is currently in press *Concerning Möbius's Function*, [Rendiconti del Circolo Matematico di Palermo] I append Littlewood's theorem: from

Riemann's hypothesis, it follows that

$$M(u) = O\left(n^{\frac{1}{2} + \frac{10 \lg \lg \lg n}{\lg \lg n}}\right).$$

(4), (5), and (6) follow from this; If Riemann's hypothesis is true, then

$$C(n) = O \sum_{a=1}^{n} \frac{1}{a} \left(\frac{n}{a}\right)^{\frac{1}{2}} n^{\frac{10 \lg \lg \lg n}{\lg \lg n}} = O\left(n^{\frac{1}{2} + \frac{10 \lg \lg \lg n}{\lg \lg n}}\right),$$

$$I = O \sum_{r=1}^{n} \frac{n}{r} n^{\frac{20 \lg \lg \lg n}{\lg \lg n}} = O\left(n^{1 + \frac{21 \lg \lg \lg n}{\lg \lg n}}\right),$$

$$\sum_{\nu=1}^{A} \eta_n u^2 = O\left(n^{-1 + \frac{21 \lg \lg \lg n}{\lg \lg n}}\right).$$

(11) follows from (10) due to (12).

## Conclusion

The final outcome of this paper is: The correct number among the relationships (1), (2), (3), (8), (9), (10), and (11) is an integral multiple of 7.

## 5.5. Neville's Search for Structure

Even more so than with the primes, almost any way one visualizes the Farey sequence one sees haunting patterns. When one tries to capture these patterns for close study using today's mathematical constructs one finds that they are just out of reach. This is not surprising since Franel showed that the Farey sits at the mathematical boundary between regularity and randomness, the Riemann hypothesis.

On the randomness side of this boundary are the many applications that regard the sequence as decidedly irregular, in that sense that it is different from any sequence that is regular. In some cases the application is looking for regularity and doesn't want to be fooled by finding a regularity in its own search filters or by missing a regularity that is masked by regularity in these filters.

On the other side of this boundary are the applications that regard the sequence as for all practical purposes regular. Neville was a strong proponent of this point of view. Neville's stand on the essential uniformity of the Farey sequence reflects an operational belief in the Farey sequence variant of the Riemann hypothesis. He says in the Introduction to [186], "With the theoretical basis of this uniformity of density, and with its importance in applications to analytical number-theory, we are not concerned here." so it is highly likely he was familiar with the work of Franel and Landau.

Figures 5.7 and 5.8 are a table that Neville put in the Introduction to "Farey Series of 1025" to make his point about the regularity of the series. Of this table he says "...we cannot but feel that the general uniformity of density is almost uncanny." Neville repeats the use of the adjective *uncanny* in his Farey structure paper: "...for random fractions, the accuracy of the (uniform density) rule is uncanny."

In both of the Introductions to his Royal Society tables, Neville describes in detail how to do linear interpolation on the Farey sequence. At the same time he was constructing these tables and writing these introductions, Neville published a short, terse paper entitled "The Structure of Farey Series" in which he carefully describes the nature of the regularity and uniformity he finds in the Farey sequence. In fact it was in the process of doing research for this paper

DECIMAL INDEX

SHOWING THE ESTIMATED AND THE TRUE LOCATIONS OF $n/1000$

|  | 5 | 6 | 7 | 8 | 9 |
|---|---|---|---|---|---|
| 0·00 | 004·20·20 − 43 | 005·16·20 − 21 | 006·12·19 − 14 | 007·08·19 − 18 | 008·04·19 − 13 |
| ·01 | 012·20·17 − 4 | 013·16·17 − 11 | 014·12·17 − 14 | 015·08·17 − 10 | 016·04·17 − 8 |
| ·02 | 020·20·15 − 5 | 021·16·15 − 6 | 022·12·15 + 2 | 023·08·14 − 9 | 024·04·14 − 10 |
| ·03 | 028·20·13 | 029·16·13 − 7 | 030·12·13 + 7 | 031·08·12 − 5 | 032·04·12 − 4 |
| ·04 | 036·20·10 − 3 | 037·16·10 − 7 | 038·12·10 − 1 | 039·08·10 − 3 | 040·04·01 + 1 |
| 0·05 | 044·20·08 + 2 | 045·16·08 | 046·12·08 − 3 | 047·08·07 − 5 | 048·04·07 − 8 |
| ·06 | 052·20·06 − 5 | 053·16·05 − 3 | 054·12·05 − 3 | 055·08·05 − 2 | 056·04·05 − 15 |
| ·07 | 060·20·03 − 1 | 061·16·03 − 1 | 062·12·03 − 26 | 063·08·03 − 1 | 064·04·02 − 5 |
| ·08 | 068·20·01 + 1 | 069·16·01 | 070·11·20 − 13 | 071·07·20 | 072·03·20 |
| ·09 | 076·19·19 − 2 | 077·15·18 − 1 | 078·11·18 − 3 | 079·07·18 − 4 | 080·03·18 − 3 |
| 0·10 | 084·19·16 − 2 | 085·15·16 − 4 | 086·11·16 − 3 | 087·07·16 − 1 | 088·03·15 + 2 |
| ·11 | 092·19·14 + 1 | 093·15·14 | 094·11·13 + 2 | 095·07·13 − 2 | 096·03·13 |
| ·12 | 100·19·11 − 1 | 101·15·11 − 5 | 102·11·11 − 8 | 103·07·11 − 3 | 104·03·11 + 7 |
| ·13 | 108·19·09 − 2 | 109·15·09 − 2 | 110·11·09 − 4 | 111·07·08 − 4 | 112·03·08 |
| ·14 | 116·19·07 − 2 | 117·15·07 − 2 | 118·11·06 | 119·07·06 | 120·03·06 − 2 |
| 0·15 | 124·19·04 | 125·15·04 | 126·11·04 − 3 | 127·07·04 − 8 | 128·03·03 − 2 |
| 16 | 132·19·02 | 133·15·02 + 5 | 134·11·02 − 18 | 135·07·01 − 4 | 136·03·01 + 2 |
| ·17 | 140·18·20 − 3 | 141·14·19 − 3 | 142·10·19 − 6 | 143·06·19 | 144·02·19 + 1 |
| ·18 | 148·18·17 + 3 | 149·14·17 + 2 | 150·10·17 − 2 | 151·06·17 − 2 | 152·02·16 − 1 |
| 19 | 156·18·15 + 2 | 157·14·15 + 1 | 158·10·15 + 1 | 159·06·14 + 4 | 160·02·14 + 11 |
| 0·20 | 164·18·13 + 1 | 165·14·12 − 2 | 166·10·12 | 167·06·12 | 168·02·12 + 1 |
| ·21 | 172·18·10 − 2 | 173·14·10 − 2 | 174·10·10 − 2 | 175·06·10 − 2 | 176·02·09 − 2 |
| ·22 | 180·18·08 − 5 | 181·14·08 − 6 | 182·10·07 − 8 | 183·06·07 − 6 | 184·02·07 − 3 |
| ·23 | 188·18·06 − 4 | 189·14·05 − 6 | 190·10·05 − 6 | 191·06·05 | 192·02·05 − 6 |
| ·24 | 196·18·03 − 5 | 197·14·03 − 1 | 198·10·03 − 2 | 199·06·02 + 1 | 200·02·02 + 6 |
| 0·25 | 204·18·01 − 2 | 205·14·01 − 5 | 206·09·20 − 5 | 207·05·20 | 208·01·20 − 6 |
| ·26 | 212·17·18 − 6 | 213·13·18 − 4 | 214·09·18 − 7 | 215·05·18 − 5 | 216·01·18 − 4 |
| ·27 | 220·17·16 − 4 | 221·13·16 − 6 | 222·09·16 − 4 | 223·05·15 − 3 | 224·01·15 + 1 |
| ·28 | 228·17·14 + 4 | 229·13·14 − 16 | 230·09·13 − 3 | 231·05·13 − 2 | 232·01·13 − 1 |
| ·29 | 236·17·11 − 1 | 237·13·11 + 1 | 238·09·11 + 1 | 239·05·11 | 240·01·10 + 2 |
| 0·30 | 244·17·09 − 2 | 245·13·09 − 1 | 246·09·09 | 247·05·08 − 3 | 248·01·08 − 1 |
| ·31 | 252·17·07 | 253·13·06 | 254·09·06 + 2 | 255·05·06 + 1 | 256·01·06 + 1 |
| ·32 | 260·17·04 + 2 | 261·13·04 + 3 | 262·09·04 + 1 | 263·05·04 + 4 | 264·01·04 + 5 |
| ·33 | 268·17·02 − 13 | 269·13·02 − 5 | 270·09·01 − 5 | 271·05·01 + 1 | 272·01·01 − 8 |
| ·34 | 276·16·20 | 277·12·19 + 2 | 278·08·19 − 1 | 279·04·19 − 1 | 279·20·19 − 2 |
| 0·35 | 284·16·17 + 1 | 285·12·17 + 2 | 286·08·17 + 7 | 287·04·17 | 287·20·16 − 7 |
| ·36 | 292·16·15 + 3 | 293·12·15 + 1 | 294·08·14 + 2 | 295·04·14 + 2 | 293·20·14 + 1 |
| ·37 | 300·16·13 + 2 | 301·12·12 | 302·08·12 | 303·04·12 + 2 | 303·20·12 + 2 |
| ·38 | 308·16·10 | 309·12·10 + 1 | 310·08·10 + 1 | 311·04·09 + 2 | 311·20·09 − 3 |
| 39 | 316·16·08 + 3 | 317·12·08 + 1 | 318·08·07 + 1 | 319·04·07 + 4 | 319·20·07 + 9 |
| 0·40 | 324·16·05 + 1 | 325·12·05 + 2 | 326·08·05 + 1 | 327·04·05 − 2 | 327·20·04 + 3 |
| ·41 | 332·16·03 + 3 | 333·12·03 + 3 | 334·08·03 − 1 | 335·04·02 + 2 | 335·20·02 + 1 |
| ·42 | 340·16·01 | 341·11·20 + 2 | 342·07·20 + 4 | 343·03·20 + 7 | 343·19·20 − 8 |
| ·43 | 348·15·18 − 1 | 349·11·18 + 1 | 350·07·18 + 2 | 351·03·18 + 2 | 351·19·17 + 9 |
| ·44 | 356·15·16 − 3 | 357·11·16 − 1 | 358·07·16 + 1 | 359·03·15 + 3 | 359·19·15 − 5 |
| 0·45 | 364·15·14 − 5 | 365·11·13 − 1 | 366·07·13 + 2 | 367·03·13 + 1 | 367·19·13 + 4 |
| ·46 | 372·15·11 + 6 | 373·11·11 + 4 | 374·07·11 + 2 | 375·03·11 + 3 | 375·19·10 + 2 |
| ·47 | 380·15·09 | 381·11·09 + 3 | 382·07·08 + 4 | 383·03·08 + 3 | 383·19·08 + 2 |
| ·48 | 388·15·07 + 3 | 389·11·06 + 5 | 390·07·06 + 2 | 391·03·06 + 2 | 391·19·06 + 3 |
| ·49 | 396·15·04 + 7 | 397·11·04 + 9 | 398·07·04 + 11 | 399·03·03 + 22 | 399·19·03 + 50 |

CRITICAL TABLE: TERM ADJUSTMENT

| $n$/320 | 0 | 1 | 2 | 3 | 4 | 5 | 6 | 7 | 8 | 9 |
|---|---|---|---|---|---|---|---|---|---|---|
|  | 0·000 | 0·003 | 0·006 | 0·009 | 0·012 | 0·016 | 0·019 | 0·022 | 0·025 | 0·028 |

| $n$/320 | 10 | 11 | 12 | 13 | 14 | 15 | 16 | 17 | 18 | 19 |
|---|---|---|---|---|---|---|---|---|---|---|
|  | 0·031 | 0·034 | 0·037 | 0·041 | 0·044 | 0·047 | 0·050 | 0·053 | 0·056 | 0·059 |

FIGURE 5.7. Decimal Index of $F_{1025}$

that he discovered that Goodwyn's tables were incomplete and that he would have to build his own. The result of this side trip was as we know now was *The Farey Series of Order* 1025, *Displaying Solutions of the Diophantine Equation* $bx - ay = 1$.

DECIMAL INDEX

SHOWING THE ESTIMATED AND THE TRUE LOCATIONS OF $n/1000$

| | 0 | 1 | 2 | 3 | 4 |
|---|---|---|---|---|---|
| 0·00 | 001·01·01 | 001·17·01 − 204 | 002·13·01 − 101 | 003·08·20 − 60 | 004·04·20 − 43 |
| ·01 | 008·20·19 − 14 | 009·16·18 − 22 | 010·12·18 − 12 | 011·08·18 − 17 | 012·04·18 − 4 |
| ·02 | 016·20·16 − 1 | 017·16·16 − 2 | 018·12·16 − 8 | 019·08·16 − 9 | 020·04·15 − 5 |
| ·03 | 024·20·11 − 9 | 023·16·14 − 4 | 026·12·13 − 1 | 027·08·13 + 1 | 028·04·13 |
| ·04 | 032·20·12 − 10 | 033·16·11 − 3 | 034·12·11 − 9 | 035·08·11 − 1 | 036·04·11 − 5 |
| 0·05 | 040·20·09 + 1 | 041·16·09 + 1 | 042·12·09 − 1 | 043·08·08 + 2 | 044·04·08 + 4 |
| ·06 | 048·20·07 − 3 | 049·16·07 − 3 | 050·12·06 + 1 | 051·08·06 − 4 | 052·04·06 |
| ·07 | 056·20·04 − 4 | 057·16·04 − 3 | 058·12·04 − 5 | 059·08·04 − 3 | 060·04·04 + 2 |
| ·08 | 064·20·02 − 2 | 065·16·02 − 1 | 066·12·02 | 067·08·01 + 3 | 068·04·01 − 1 |
| ·09 | 072·19·20 + 1 | 073·15·20 − 26 | 074·11·19 | 075·07·19 + 7 | 076·03·19 − 1 |
| 0·10 | 080·19·17 − 5 | 081·15·17 − 3 | 082·11·17 − 1 | 083·07·17 − 4 | 084·03·16 − 5 |
| ·11 | 088·19·15 | 089·15·45 + 32 | 090·11·15 − 3 | 091·07·14 − 2 | 092·03·14 − 1 |
| ·12 | 096·19·13 − 1 | 097·15·12 | 098·11·12 − 5 | 099·07·13 − 3 | 100·03·12 − 3 |
| ·13 | 104·19·10 − 3 | 105·15·10 − 2 | 106·11·10 − 2 | 107·07·10 + 1 | 108·03·09 − 3 |
| ·14 | 112·19·08 | 113·15·08 + 3 | 114·11·07 + 3 | 115·07·07 − 41 | 116·03·07 − 2 |
| 0·15 | 120·19·06 + 1 | 121·15·05 − 1 | 122·11·05 − 2 | 123·07·05 − 1 | 124·03·05 − 9 |
| ·16 | 128·19·03 − 1 | 129·15·03 − 2 | 130·11·03 − 1 | 131·07·03 − 3 | 132·03·02 |
| ·17 | 136·19·01 − 1 | 137·15·01 − 2 | 138·10·20 − 1 | 139·06·20 − 2 | 140·02·20 − 7 |
| ·18 | 144·18·19 | 145·14·18 + 2 | 146·10·18 − 8 | 147·06·18 + 1 | 148·02·18 |
| ·19 | 152·18·16 | 153·14·16 + 2 | 154·10·16 + 1 | 155·06·15 − 4 | 156·02·15 + 2 |
| 0·20 | 160·18·14 + 1 | 161·14·14 − 6 | 162·10·13 | 163·06·13 + 3 | 164·02·13 + 2 |
| ·21 | 168·18·11 + 3 | 169·14·11 + 5 | 170·10·11 + 1 | 171·06·11 | 172·02·10 + 3 |
| ·22 | 176·18·09 − 4 | 177·14·09 − 4 | 178·10·09 + 6 | 179·06·08 − 5 | 180·02·08 − 3 |
| ·23 | 184·18·07 − 6 | 185·14·06 − 10 | 186·10·06 − 5 | 187·06·06 − 4 | 188·02·06 − 5 |
| ·24 | 192·18·04 − 2 | 193·14·04 − 3 | 194·10·04 − 3 | 195·06·04 − 4 | 196·02·03 − 4 |
| 0·25 | 200·18·02 − 4 | 201·14·02 − 13 | 202·10·02 − 8 | 203·06·01 − 3 | 204·02·01 − 7 |
| ·26 | 208·17·20 − 7 | 209·13·19 − 8 | 210·09·19 − 6 | 211·05·19 − 2 | 212·01·19 − 6 |
| ·27 | 216·17·17 − 4 | 217·13·17 − 3 | 218·09·17 − 3 | 219·05·17 − 10 | 220·01·16 − 3 |
| ·28 | 224·17·15 | 225·13·15 − 3 | 226·09·14 + 1 | 227·05·14 + 5 | 228·01·14 − 1 |
| ·29 | 232·17·13 − 1 | 233·13·12 | 234·09·12 | 235·05·12 − 1 | 236·01·12 + 4 |
| 0·30 | 240·17·10 + 1 | 241·13·10 | 242·09·10 | 243·05·09 + 7 | 244·01·09 − 2 |
| ·31 | 248·17·08 + 1 | 249·13·08 + 1 | 250·09·07 + 2 | 251·05·07 | 252·01·07 + 2 |
| ·32 | 256·17·05 + 5 | 257·13·05 − 1 | 258·09·05 + 2 | 259·05·05 + 1 | 260·01·05 + 3 |
| ·33 | 264·17·03 + 6 | 265·13·03 + 12 | 266·09·03 + 13 | 267·05·02 + 99 | 268·01·02 − 32 |
| ·34 | 272·17·01 | 273·13·01 − 3 | 274·08·20 | 275·04·20 − 3 | 275·20·20 − 1 |
| 0·35 | 280·16·18 + 1 | 281·12·18 | 282·08·18 | 283·04·18 − 17 | 283·20·17 + 2 |
| ·36 | 288·16·16 + 1 | 289·12·16 + 2 | 290·08·16 + 4 | 291·04·15 + 6 | 291·20·15 − 1 |
| ·37 | 296·16·14 + 2 | 297·12·13 + 2 | 298·08·13 + 4 | 299·04·13 + 3 | 299·20·13 + 3 |
| ·38 | 304·16·11 + 3 | 305·12·11 − 11 | 306·08·11 + 2 | 307·04·11 − 4 | 307·20·10 + 4 |
| ·39 | 312·16·09 + 3 | 313·12·09 + 3 | 308·08·08 + 3 | 313·04·08 | 315·20·08 + 1 |
| 0·40 | 320·16·07 + 1 | 321·12·06 − 6 | 322·08·06 | 323·04·06 − 2 | 323·20·06 + 2 |
| ·41 | 328·16·04 + 1 | 329·12·04 | 330·08·04 − 2 | 331·04·04 + 3 | 331·20·03 + 2 |
| ·42 | 336·16·02 | 337·12·01 + 14 | 338·08·01 + 2 | 339·04·01 + 6 | 336·20·01 + 2 |
| ·43 | 344·15·20 | 345·11·19 + 2 | 346·07·19 + 1 | 347·03·19 | 347·19·19 − 1 |
| ·44 | 352·15·17 + 2 | 353·11·17 + 1 | 354·07·17 + 1 | 355·03·16 + 3 | 355·19·16 + 6 |
| 0·45 | 360·15·15 + 1 | 361·11·14 − 3 | 362·07·14 | 363·03·14 − 2 | 363·19·14 |
| ·46 | 368·15·12 + 2 | 369·11·12 + 2 | 370·07·13 + 1 | 371·03·12 | 371·19·11 + 4 |
| ·47 | 376·15·10 + 4 | 377·11·10 + 3 | 378·07·10 + 2 | 379·03·09 + 1 | 379·19·09 + 1 |
| ·48 | 384·15·08 + 5 | 385·11·07 + 6 | 386·07·07 + 3 | 387·03·07 − 1 | 387·19·07 |
| ·49 | 392·15·05 + 1 | 393·11·05 + 5 | 394·07·05 + 6 | 395·03·05 | 395·19·04 + 6 |
| 0·50 | 400·15·03 | | | | |

| | | CRITICAL TABLE: LINE ADJUSTMENT | | | | | |
|---|---|---|---|---|---|---|---|
| $n$ | 0 | 1 | 2 | 3 | 4 | 5 | 6 | 7 |
| $n/16$ | 0·000 | 0·062 | 0·125 | 0·187 | 0·250 | 0·312 | 0·375 | 0·437 |
| $n$ | 8 | 9 | 10 | 11 | 12 | 13 | 14 | 15 |
| $n/16$ | 0·500 | 0·562 | 0·625 | 0·687 | 0·750 | 0·812 | 0·875 | 0·937 |

FIGURE 5.8. Decimal Index of $F_{1025}$

It was when I had begun, late in 1945, to study the numerical structure of Farey series, and was hoping to learn of a series actually written out, high enough in order for the characteristic properties to be exhibited clearly, that I read in the Guide (p. 8) that the series of order 1000 is to be found in Henry Goodwyn's A Tabular Series of Decimal Quotients ... (London, 1823), a volume of v+ 153 pages. The bibliographer should have been suspicious; the series of order 1000 consists of 304193 terms, and the implication that 2000 fractions with decimal equivalents to eight places are printed on a page is enough to throw doubt on any assertion. As it happens, I have handled Goodwyn's exquisite little book. I knew that the elaborate title quoted in the Guide (p. I03) describes a project, not a performance, an aspiration, not an accomplishment. What was published in 1823 was a first installment, with the promise that more would follow if the enterprise received adequate support. In fact no more was published, and although Goodwyn left a mass of papers, no one knows what became of them, nor is there a record of their contents. [**185**]

The Guide to which Neville refers is Lehmer's *Guide to Tables in the Theory of Numbers*.

In his "The Structure of the Farey Series" paper Neville states the following theorem:

THEOREM 7. *The density of a Farey series is roughly uniform.*

He knows full well that this is not in any sense the statement of a real mathematical theorem nor does he offer a proof a real mathematical proof. But the statement of the observation as a theorem does reveal his strong conviction as to its truth, as least for as he says "practical purposes." Most of the other theorems in the paper are real theorems with real proofs but Theorem 7 is just out of his reach too.

A metric that Neville uses throughout the structure paper is

$$|vx - uy|$$

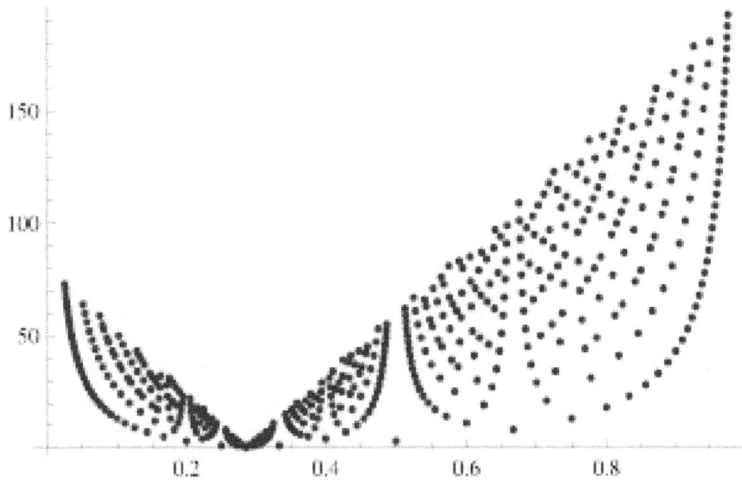

FIGURE 5.9. $|vx - uy|$ for $\frac{2}{7}$ and $F_{40}$

where $\frac{u}{v}$ is a fixed fraction and $\frac{x}{y}$ ranges over a Farey sequence. Figure 5.9 is a plot of this metric for $\frac{u}{v} = \frac{2}{7}$ and $\frac{x}{y}$ ranging over $F_{40}$. You can almost hear the siren song of regularity.

## 5.6. Capturing Regularization

The core of the relationship between the Farey sequence and the Riemann hypothesis is that the Farey sequence becomes a sufficiently uniform sequence pretty quickly. This is exactly the observation that Neville had noted and come back to again and again in his mathematical tables and in his papers on the Farey sequence.

Seemingly, a link between the Farey sequence and the Riemann hypothesis could be constructed by providing precise definitions of "sufficiently" and "pretty quickly". For "sufficiently" Franel and Landau compute a distance between a Farey sequence and a fixed uniform sequence and for "pretty quickly" they say how fast this distance grows. There are many sequences that become uniform and there are many ways of computing a distance between any two of these so this would seem to be a fertile row to hoe, at least for amateur mathematicians.

The approach of Franel and Landau is what Fowler would call arithmetized because they model regularity as computed distance and, more tellingly, treat the elements of the Farey sequence as fractions.

In the spirit of Fowler one is led to wonder if there is a non-arithmetic statement of the Riemann hypothesis, perhaps even one based on the mediant. For example, suppose one conjectured that the Riemann hypothesis was true if and only if the mean and the mediant converge. This is just another way of saying that "the Farey sequence becomes a sufficiently uniform sequence pretty quickly" but it is a non-arithmetic way. Readers who would like to pursue this line of speculation are referred to [152].

Non-arithmetics aside, here is an example of approaching the problem of using the Farey sequence to prove the Riemann hypothesis through the back door rather than the front door. We posit a measure of convergence to regularity, estimate its rate of convergence as the order of the Farey sequence goes to infinity and then see if we can link our measure to a statement of the Riemann hypothesis.

As our element-level measure of regularity rather than computing the distance between a Farey fraction and a fraction in a uniform sequence as Franel and Landau did we will compute the difference between the mean and the mediant of adjacent elements of the Farey sequence. In other words, if $\frac{p_i}{q_i}$ and $\frac{p_{i+1}}{q_{i+1}}$ are adjacent elements of $F_m$, the Farey sequence of order $m$, then

$$\delta_m(i) = \frac{\frac{p_{i+1}}{q_{i+1}} + \frac{p_i}{q_i}}{2} - \frac{p_i + p_{i+1}}{q_i + q_{i+1}}$$

Figure 5.10 displays $\delta_m(i)$ for each adjacent pair in the Farey sequence of order 100. Figure 5.11 plots the sum of the absolute value of these differences across all the adjacent pairs in $F_m$ for $m = 10(10)200$.

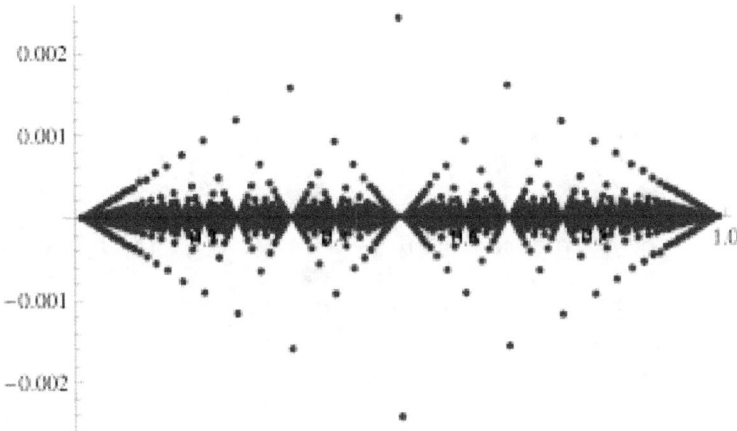

FIGURE 5.10.   Difference between mean and mediant for $m = 50$

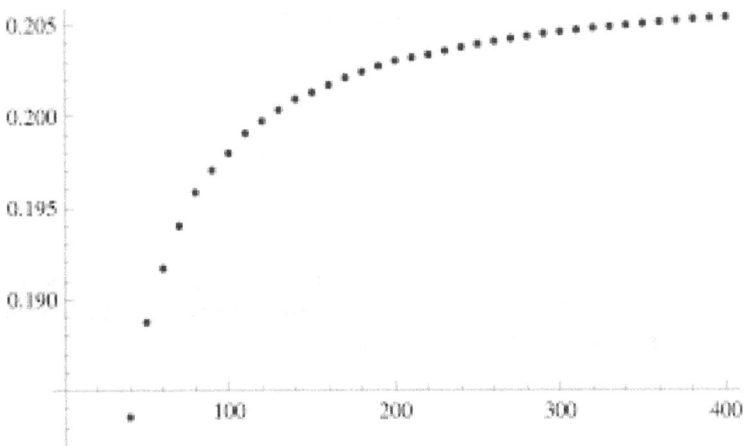

FIGURE 5.11.   Distance between mean and mediant for $m = 10(10)200$

$\delta_m(i)$ goes to zero as $m$ goes to infinity simply because the distance between adjacent elements goes to zero so what we are really interested in is the

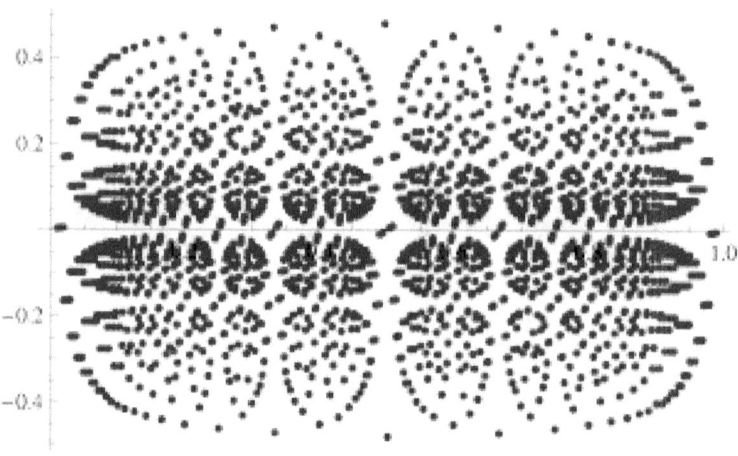

FIGURE 5.12.    Relative term-wise difference $m = 100$

convergence of the mean and the mediant relative to the distance between the adjacent elements:

$$\hat{\delta}_m(i) = \frac{\frac{\frac{p_i}{q_i} + \frac{p_{i+1}}{q_{i+1}}}{2} - \frac{p_i + p_{i+1}}{q_i + q_{i+1}}}{\frac{p_{i+1}}{q_{i+1}} - \frac{p_i}{q_i}}$$

Figure 5.12 displays $\hat{\delta}_m(i)$ for each adjacent pair in the Farey sequence of order 100. Figure 5.13 plots the sum of the absolute value of these differences across all the adjacent pairs in $F_m$ for $m = 10(10)200$.

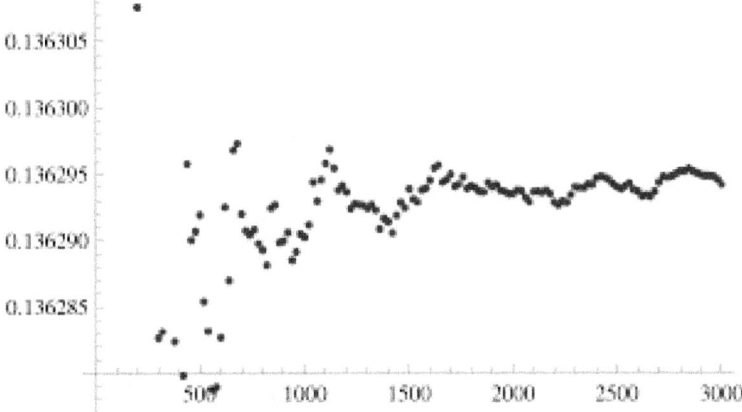

FIGURE 5.13.   Sum of the relative distances for $m = 10(10)200$

CHAPTER 6

# Explorations and Peregrinations

The Franel-Landau work provides amateur mathematicians with an accessible entry point to the avocational study of the Riemann hypothesis. If the suspicions of some professional mathematicians are correct that cracking the hypothesis will require some wholly new mathematical machinery, then history says that amateurs in unconstrained wandering have as good a shot at stumbling across this new machinery as do the professionals.

To get hold of the behavior of the Franel-Landau as a function of $m$ one has to be able work with $F_m$ as a function of $m$ just as they did. One way to accomplish this is to come up with a closed expression for the $i^{th}$ term of $F_m$ as a function of $i$ and $m$. This however is equivalent to building an algorithm to generate primes and this road is well-traveled ([**220**], [**97**]).

## 6.1. The Integer Part Function

The integer part function is another motif of mathematics. It is found frequently in number theory, where it is used to economically describe numerical relationships and properties. While the working description of the integer part function – just throw away everything after the decimal point – is immediately grasped, the function is surprisingly difficult to use analytically. The integer part function is also referred to as the floor function and the greatest integer function.

The integer part function $[x]$ of a real number $x$ is defined as that integer $n$ such that

$$0 \leq sign(x)(x - n) < 1$$

The part that is thrown away is called the fractional part, $\langle x \rangle$

$$\langle x \rangle = x - [x]$$

where for the sake of completeness at integer values $n$ is customary to define

$$\lim_{\epsilon \to 0^+} \langle n + \epsilon \rangle = 0$$

and

$$\lim_{\epsilon \to 0^+} \langle n - \epsilon \rangle = 1$$

As a relevant example of the utility of the integer part function, the Farey sequence of order $m$, $\{n_i/d_i\}$ can be described in terms of the integer part function as follows. For $d_0 = 1$ and $d_1 = m$ set

$$d_{i+1} = \left[ \frac{m + d_{i-1}}{d_i} \right] d_i - d_{i-1},$$

and for $n_0 = 0$ and $n_1 = 1$ set

$$n_{i+1} = \left[ \frac{m + d_{i-1}}{d_i} \right] n_i - n_{i-1},$$

for $i$ up to $\sum_{k=1}^{m} \phi(k)$ where $\phi(k)$ is the Euler totient function; viz. the number of integers less than equal to $k$ that are coprime to $k$.

One way of getting a closed form for the Farey sequence as a function, of $m$ is to come up with a closed expression for the $i^{th}$ term of the series as a function of $m$. For example, the $3^{rd}$ term as a function of $m$ is

$$F_3^m = \frac{-1 + \left[1 + \frac{1}{m}\right] \left[\frac{2m}{-1 + m\left[1 + \frac{1}{m}\right]}\right]}{-m + \left(-1 + m\left[1 + \frac{1}{m}\right]\right) \left[\frac{2m}{-1 + m\left[1 + \frac{1}{m}\right]}\right]}$$

We can observe that for $m > 1$

$$\left[1 + \frac{1}{m}\right] = 1$$

and

$$\left[\frac{2m}{-1 + m\left[1 + \frac{1}{m}\right]}\right] = 2$$

but starting with

$$F_4^m = \frac{-1 + \left[\frac{2m-1}{m-2}\right]}{1 - m + (m-1)\left[\frac{2m-1}{m-2}\right]}$$

we're in an analytical hole.

One way out of this hole might be to use a purely analytical expression for the integer part function of which there are quite a few. For example, Equation 6.1.1 is way of defining the integer part function that uses the tangent and cotangent functions.

$$[x] = x + \frac{\arctan(\cot(\pi x))}{\pi} - \frac{1}{2} \tag{6.1.1}$$

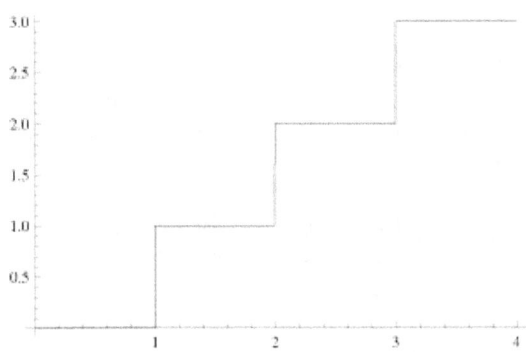

FIGURE 6.1. $[x]$ using the tangent and cotangent functions

Alternatively equation 6.1.2 is way of defining the integer part function that uses the sine function.

$$[x] = x + \sum_{i=1}^{\infty} \frac{\sin(2\pi i x)}{k} - \frac{1}{2} \tag{6.1.2}$$

Finally, equation 6.1.3 is a way of approximating the integer part function that uses the cosine function.

$$d(x) = \frac{1}{4} - \frac{1}{\pi} \sum_{n=0}^{\infty} \frac{\cos(2\pi(n+1)x)}{(2n+1)^2} \tag{6.1.3}$$

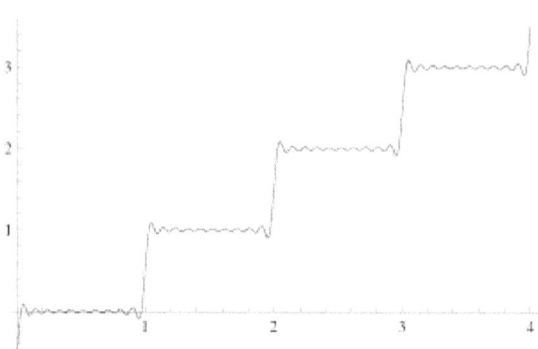

FIGURE 6.2. $[x]$ using the sine function

$d(x)$ is the distance between $x$ and the nearest integer. It is increasing when the nearest integer is below $x$ and decreasing when the nearest integer is above $x$. Equation 6.1.4 uses this property of $d(x)$ to build the rather fanciful approximation to $[x]$ shown in Figure 6.3.

$$[x] \approx \begin{cases} x - d(x) & \text{for } d'(x) > 0 \\ x + d(x) - 1 & \text{for } d'(x) <= 0 \end{cases} \qquad (6.1.4)$$

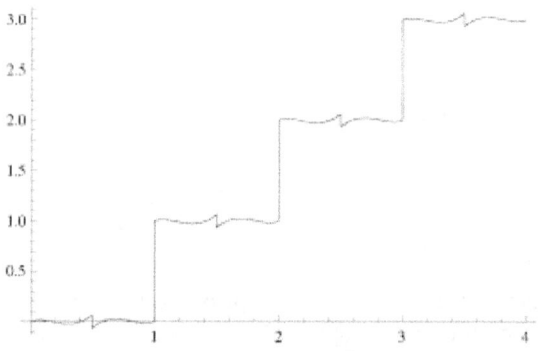

FIGURE 6.3. Approximating $[x]$ using the cosine function

At first blush, none of these series would seem to be of much help in cracking 3.7 regardless of what 3.7 might look like an arbitrary value of $m$.

A book published by Folke Ryde in 1973 [211], *Aspects of the greatest integer function – Part I*, built a theoretical framework for the greatest integer function but the approach hasn't gained widespread use.

In the context of studying computer science's 4/3 problem, Håkan Lennerstad and Lars Lundberg published a technical report in 2006 entitled "Generalizations of the floor and ceiling functions using the Stern-Brocot tree" [165] that provides a more tractable handle on the integer part function and more importantly for our purposes, a practical and compelling application of the mediant function.

> Assume that $a$ is an integer and $b$ is a positive integer. A decomposition
> $$\frac{a}{b} = \frac{a_1 + a_2 + \ldots + a_c}{b_1 + b_2 + \ldots + b_c}$$
> is **uniform from below** if there is no other decomposition with a larger
> $$\lfloor \frac{a}{b} \rfloor_c = \min_i \frac{a_i}{b_i}.$$
> A decomposition is **uniform from above** if there is no other decomposition with a smaller
> $$\lceil \frac{a}{b} \rceil_c = \max_i \frac{a_i}{b_i}.$$
> A decomposition is **uniform** if it is both uniform from below and uniform from above.

$\lfloor \frac{a}{b} \rfloor_c$ is the Lennerstad/Lundberg generalization of the floor or integer part function and $\lceil \frac{a}{b} \rceil_c$ is their generalization of the ceiling function. Since the $b_i$ are positive integers, these two functions are defined for values of $c$ between 1 and $b$. When $c$ is one it's obvious that
$$\lfloor \frac{a}{b} \rfloor_1 = \lfloor \frac{a}{b} \rfloor_1 = \frac{a}{b}$$

The functions can be thought of as generalizations of floor and ceiling respectively because when $c$ is equal to $b$

$$\lfloor \frac{a}{b} \rfloor_b = \lfloor \frac{a}{b} \rfloor \quad \text{and} \quad \lceil \frac{a}{b} \rceil_b = \lceil \frac{a}{b} \rceil$$

By way of an example, the 7 values of $\lfloor \frac{4}{7} \rfloor_c$ are

$$\left\{ \frac{4}{7}, \frac{1}{2}, \frac{1}{2}, \frac{1}{2}, 0, 0, 0 \right\}$$

and the 7 values of $\lfloor \frac{4}{7} \rfloor_c$

$$\left\{ \frac{4}{7}, \frac{3}{5}, \frac{2}{3}, 1, 1, 1, 1 \right\}$$

## 6.2. Mediant Factorization

In their technical report Lennerstad and Lundberg show that for each $c$ there is a unique uniform decomposition for any fraction that we could justifiably think of as the *mediant factorization* of the fraction. They furthermore prove that this unique decomposition contains at most three distinct ratios. If we know the minimum of these three values, $\lfloor \frac{a}{b} \rfloor_c$ and the maximum of the three values, $\lceil \frac{a}{b} \rceil_c$ then it's rather easy to figure out how many ratios there are as well as their value(s).

Returning to the above example, the unique 3-part factorization $\frac{4}{7}$ uses two ratios:

$$\frac{4}{7} = \frac{1+1+2}{2+2+3}$$

To bring this all back to the Brocot-Stern tree, Lennerstad and Lundberg give the following $O(b)$ algorithm for computing $\lfloor \frac{a}{b} \rfloor_c$ and $\lceil \frac{a}{b} \rceil_c$:

```
GeneralizedFloorAndCeiling[ax_,bx_]:=Module[{a,b,n,d,
A={{0,1},{1,1}},G={0,1},l=0,L=1,r=1,R=1,g=2,p=2,F={},C={},
x,y,q,u,v,i, floor,ceiling},
a=ax;b=bx;
(* Compute the Stern-Brocot tree *)
d=GCD[a,b];a =a/d;b=b/d;a = Mod[a,b];
```

```
While[True,
   A=Insert[A,{l+r,L+R},p];G=Insert[G, g,p];
   g++;
   If[a/b==(l+r)/(L+R),Break[]];
   If[a/b<(l+r)/(L+R),r=l+r;R=L+R;Continue[]];
   If[a/b>(l+r)/(L+R),l=l+r;L=L+R;p++;Continue[]];
];
(* Stern-Brocot tree to compute the floors and ceilings *)
x = p-1;y=p+1;u=v=d;q=g;
F=AppendN[F,{a,b},d];
C=AppendN[C,{a,b},d];
While[True,
   q--;
   i=Position[G, q][[1,1]];
   If[q==0,
      F=AppendN[F,A[[1]],u-v];
      Break[];
   ];
   If[i<p,
      F=AppendN[F,A[[x]],u];
      x--;v += u;Continue[];
   ];
   If[i>p,
      C=AppendN[C,A[[y]],v];
      y++;u +=v;Continue[];
   ];
];
floor=Table[F[[i,1]]/F[[i,2]],{i,1,Length[F]}];
ceiling=Table[C[[i,1]]/C[[i,2]],{i,1,Length[C]}];
{floor, ceiling}
]
```

## 6.3. The Mayer-Erdös Constant

In a short, three and one-half page paper entitled "A Note On Farey Series" Paul Erdös [**67**] proved the following generalization of a result of A.E. Mayer [**176**]:

THEOREM 3. *There exists an absolute constant c such that, if $m > ck$, then for elements $a/b$ of $F_m$*

$$(a_{k+x} - a_x)(b_{k+x} - b_x) \geq 0$$

Two fractions that satisfy the condition of the theorem are said to be *similarly ordered*, so named because the numerators of two exhibit the same ordering as the denominators. Thus, $\frac{1}{2}$ and $\frac{2}{3}$ are similarly ordered but $\frac{2}{5}$ and $\frac{3}{4}$ are not.

Erdös proof sets an upper bound on $c$ of 192. At the end of his proof Erdös says "I have not been able to find the best possible value of the constant $c$ in the above result."

It is not true that if all Farey fractions $k$ apart in $F_m$ are similarly ordered that this property will hold for $m' > m$ so once we find an $m$ with no similarly-ordered violations we aren't done. The code below investigates $c$ by taking as a parameter the number of sequential $m$ values that show no similarly-ordered violations as the criteria for finding $c$ for a particular value of $k$.

```
FareySequence[n_] :=
 Join[{0}, Union[Table[i/j, {j, n}, {i, j - 1}] // Flatten], {1}]

FareyLength[n_] := Sum[EulerPhi[i], {i, 1, n}] + 1

FareyStart[k_] := Module[{m = 2},
  While[FareyLength[m] < k, m++];
  m
  ]

SimilarlyOrdered[a_, b_] :=
    (Numerator[b]   - Numerator[a]) *
    (Denominator[b] - Denominator[a]) >= 0;
```

```
SimilarlyOrderedTest[k_, m_] := Module[{f, i, l},
  l = FareyLength[m];
  f = FareySequence[m];
  For[i = 1, i <= l - k, i++,
   If[!SimilarlyOrdered[f[[i]], f[[i + k]]],
     Return[False]
     ];
   ];
  Return[True];
  ]

MayerErdosConstant[k_, runLength_] := Module[{m, c = 0, s},
  For[m = FareyStart[k], True, m++,
   s = SimilarlyOrderedTest[k, m];
   c = Which[s && c != 0, c, s && c == 0 , m, True, 0];
   If[c != 0 && m - c > runLength,
    Break[];
    ];
   ];
  c/k
  ]
```

Figure 6.4 is a plot of the MayerErdosConstant with the runLength parameter set to 5.

Erdös concludes the paper by stating but not proving three additional theorems about the Farey sequence.

THEOREM 4. *To every $\epsilon > 0$ there exists a $c = c(\epsilon)$ such that any interval of length $(1 + \epsilon)/m$ contains at least $cm$ Farey fractions of order $m$.*

THEOREM 5. *If $f(m) \to \infty$ as $m \to \infty$, any interface of length $m^{-1}f(m)$ contains*

$$\frac{3}{\pi^2}mf(m) + o(mf(m))$$

*Farey fractions of order $m$.*

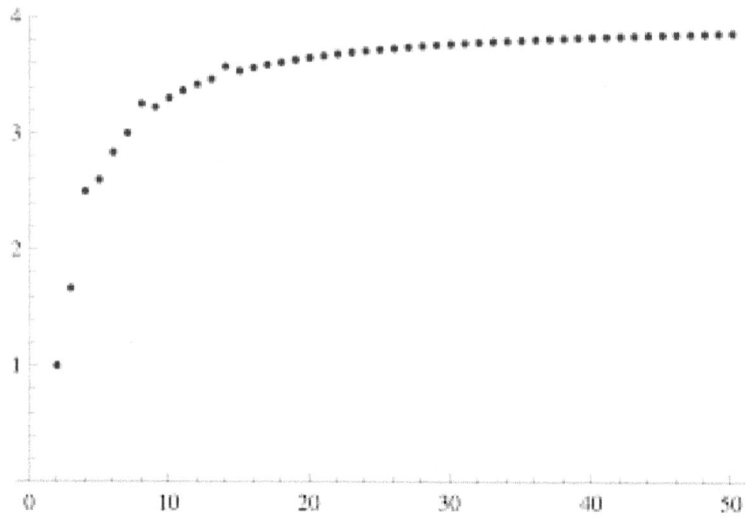

FIGURE 6.4. Mayer-Erdös Computation with Run Length of 5

THEOREM 6. *There exists a constant $c_1$ such that any interval of length $L = k^{c_1}$ contains a set of at lead $k$ mutually prime integers. Furthermore*

$$L(k) > c_2 \frac{k \log k \log \log k}{(\log \log k)^2}.$$

He says at the end of the paper "It would be interesting to have a good estimate for the best possible value $L(k)$ of $L$ from below."

## 6.4. Ocagne's Recursion

Rather than work with $F_m$ as a whole, one can try to work with individual terms of the Franel-Landau sum and try to express them as a function of $m$. There are a number of algorithms to generate the individual terms of the Farey sequence of order $m$. As above, let $n_f$ be the numerator and $d_f$ be the denominator of a rational number $f$ but write $f$ as $(n_f, d_f)$.

An engineer in the Corps des Ponts et Chaussées and the mathematician who first explored the mathematics underpinning nomograms, Maurice d'Ocagne, also studied the mediant and the Farey sequence.

In a paper in 1885 [189] he turned the mediant property of the Farey sequence into the following recursion for generating the sequence:

$$f_{i+1} = \left[ \frac{d_{f_{i-1}} + m}{d_{f_i}} \right] f_i - f_{i-1}$$

where $f_1 = (0, 1)$ and $f_2 = (1, m)$.

That a simple recursion with a two-step memory could encode the primes and prime factorization is to my thinking a shadow on Flatland.

Ocagne's recursion builds closed forms for the terms in the Farey sequence as a function for $m$ and $i$ that are expressed as compositions of the integer part function. For example, using this recursion the third term as a function of $m$ can be expressed as

$$\frac{-1 + \left[1 + \frac{1}{m}\right] \left[ \frac{2m}{-1 + m\left[1 + \frac{1}{m}\right]} \right]}{-m + \left(-1 + m\left[1 + \frac{1}{m}\right]\right) \left[ \frac{2m}{-1 + m\left[1 + \frac{1}{m}\right]} \right]}.$$

Of course since

$$[2] = 2, \quad \left[1 + \frac{1}{m}\right] \quad \text{and} \quad \left[ \frac{2m}{-1 + m} \right]$$

for $m \geq 2$ this expression reduces to

$$\frac{1}{m - 2}.$$

One can always achieve such reductions at the tails of the Farey sequence as $m$ increases. It is the middle part that is the challenge and that's where we are stuck with irreducible nesting of the integer part function.

If we let $O_i^{(k,l,m)}$ denote $i^{th}$ term of the Ocagne iteration with starting condition $f_1 = (0, k)$ and $f_2 = (1, l)$, then we have an expression and a recursion for the terms of the sum in Franel's theorem:

$$O_i^{(1,m,m)} - O_i^{(k,\Phi(m),\Phi(m))}$$

## 6.5. Primes and Twin Primes

In order to generate $F_{m+1}$ from $F_m$ let element $f = \frac{n_f}{d_f}$ of the Farey set $F_m$ define

$$\sigma_f(m) = d_f - ((m - n_f) \bmod d_f).$$

$\sigma_f(m)$ is a periodic counter that counts the number of orders from the order $m$ until a new fraction appears next to $f$.

The following three properties of $\sigma_f$ follow directly from its definition:

- $\sigma_f(d_f) = n_f$,
- If $\sigma_f(m) = c$, then $\sigma_f(m + kd_f) = c$ for $k \geq [1 - m/d_f]$
- If $\sigma_f(m) = k$, then $\sigma_f(m + k - 1) = 1$.

Using $\sigma_f$ then the Farey set $F_m$ can be expressed recursively in terms of $F_{m-1}$ as follows:

$$F_m = F_{m-1} \bigcup \{(m - d_f, m) \mid f \in F_{m-1} \text{ and } \sigma_f(m - 1) = 1\}$$

where $F_1 = \{(0, 1)\}$.

LEMMA 1. *If $\sigma_f(m - 1) = 1$, then $\sigma_f(m) = d_f$.*

PROOF. $d_f - ((m - 1 - n_f) \bmod d_f) = 1$ implies $m - n_f \bmod d_f = 0$ so $\sigma_f(m) = d_f$. $\qquad \square$

For a Farey fraction $f$, let

$$m_f(c, k) = \begin{cases} kd_f + n_f - c & c \leq n_f \\ (k + 1)d_f + n_f - c & c > n_f \end{cases}$$

$m_f(c, k)$ is the $k^{th}$ value of $m$ starting at $m = d_f$ for which $\sigma_f(m) = c$. Define

$$M_c(d) = \bigcup_{d_f = d} \{m_f(c, k), k = 0, 1, 2, \ldots\}.$$

$M_c(d)$ is the set of all $m$ such that there is a Farey fraction $f$ with denominator $d$ and $\sigma_f(m) = c$ in $F_m$. Finally, define

$$\mathcal{M}_c(d) = \bigcap_{k=1}^{d} M_c(k).$$

$\mathcal{M}_c(d)$ is the set of all $m$ such that for each $k$, $1 \leq k \leq d$, there is a Farey fraction $f$ with denominator $k$ and $\sigma_f(m) = c$ in $F_m$.

LEMMA 2. $p+1$ is a prime if and only if $p \in \mathcal{M}_1(p)$ .

PROOF. $p \in \mathcal{M}_1(p)$ if and only if there is a fraction $f$ with denominator $d$ and $\sigma_f(p) = 1$ in $F_p$ for each $d$, $1 \leq d \leq p$. In this case $|F_{p+1}| - |F_p| = p = \varphi(p+1)$ and $\varphi(p+1) = p$ if and only if $p+1$ is a prime. □

THEOREM 7. Let $p$ be a prime and set

$$\bar{p} = min\, \mathcal{M}_1(p).$$

Then $\bar{p} + 1$ is the next prime after $p$.

PROOF. Consider $d \in \mathcal{M}_1(p)$. From the definition of $\mathcal{M}_1(p)$, $1 \leq d \leq p$ and there is an $f$ with denominator $d$ in $F_{\bar{p}}$ such that $\sigma_f(\bar{p}) = 1$. If in addition $d \leq [\bar{p}/2]$ then $\sigma_f(\bar{p} - d) = 1$.

If $m = \bar{p} - d + 1$ then there is $f' \in F_m$ with $\sigma_{f'}(m) = d$ by Lemma 1. Then $\sigma_{f'}(m + d - 1) = \sigma_{f'}(\bar{p} = 1$.

To summarize, for each $m$ such that $\bar{p}/2 + 1 < m \leq \bar{p}$, there is an $f$ with denominator $m$ in $F_{\bar{p}}$ such that $\sigma_f(\bar{p}) = 1$.

Since $\bar{p}/2 < p$ by Bertrand's Postulate ([15], [38], [203], [66]) we have that there is a $f$ in $F_{\bar{p}}$ with $\sigma_f(\bar{p}) = 1$ and denominator $d$ every $d$ between 1 and $\bar{p}$. Therefore $\bar{p} \in \mathcal{M}_1(\bar{p})$ and $\bar{p} + 1$ is a prime by Lemma 2.

Since $\bar{p} = min\, \mathcal{M}_1(p)$, for each $m$, $p \leq m < \bar{p}$, there must be at least one $d$, $1 \leq d \leq m$, for which there is no fraction with denominator $d$ and $\sigma_f(m) = 1$ in $F_m$. Therefore, $m$ is not prime and $\bar{p} + 1$ is the next prime after $p$. □

The twin prime conjecture has been described in [55], [62], and [121] among many other places.

LEMMA 3. $p + 1$ *is the lesser of a prime pair if and only if*

$$p \in \mathcal{M}_1(p) \bigcap \mathcal{M}_3(p).$$

PROOF. $p$ is a prime since $p - 1 \in \mathcal{M}_1(p - 1)$.

Further since $([p/2], p, 2) \in F_p$ we have $([p/2], p, 1) \in F_{p+1}$ along with $(1, 1)$ and $(p + 1, 1)$. The condition

$$p - 1 \in \mathcal{M}_3(p - 1)$$

ensures that $(d, 1) \in F_{p+1}$ for $2 \leq d \leq p - 1$. Therefore $(d, 1) \in F_{p+1}$ for $1 \leq d \leq p + 1$. That is,

$$p + 1 \in \mathcal{M}_1(p + 1)$$

so $p + 2$ is a prime.                                                                 □

CONJECTURE 1. *Let $p$ be the lesser of a twin prime and set*

$$\bar{p} = \min\left(\mathcal{M}_1(p) \bigcap \mathcal{M}_3(p)\right).$$

*Then $\bar{p} + 1$ is the lesser of the next twin prime after $p$.*


## 6.6. The Fractional Part Function

The Ocagne recursion for the Farey sequence of a given order, $F_m$ can also be written as

$$f_{i+1} = m - \left\langle \frac{d_{f_{i-1}} + m}{d_{f_i}} \right\rangle f_i$$

Let $\langle x \rangle$ denote the fractional part of a real number $x$ where as usual for any positive integer $n$ we take

$$\lim_{\epsilon \to 0^+} \langle n + \epsilon \rangle = 0$$

and

$$\lim_{\epsilon \to 0^+} \langle n - \epsilon \rangle = 1$$

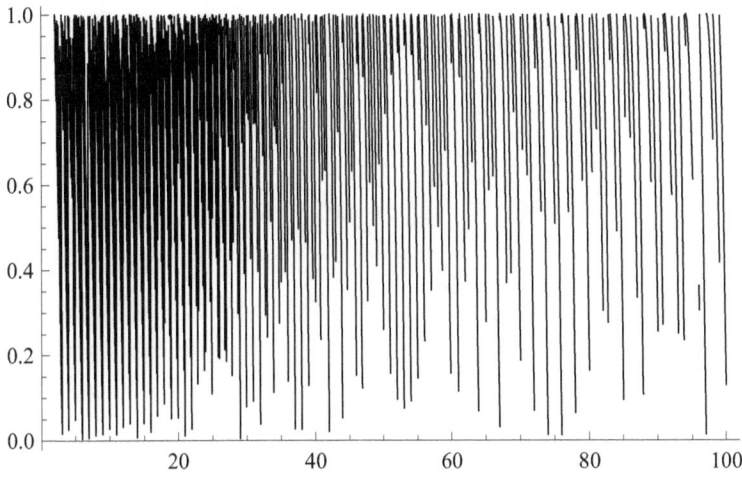

FIGURE 6.5. $f(x) = 1 - \langle x \rangle \left(1 - \langle 2813/x \rangle\right)$

A positive real number $x$ is an integer divisor of a positive integer $n$ if $\langle x \rangle$ and $\langle n/x \rangle$ are simultaneously 0 or 1. One way to arrange for this is to find $x$ increasing to an integer such that

$$\frac{1}{1 - \langle n/x \rangle} = \langle x \rangle$$

or in other words, find roots of

$$f(x) = 1 - \langle x \rangle \left(1 - \langle n/x \rangle\right)$$

Figure 6.5 is a plot of $f(x)$ for $n = 29 \times 97$.

Most root-finding algorithms, even the ones that don't require derivatives such as [**24**], fail when applied to this function. As a result we will numerically integrate the function and seek zero derivatives of the result. Depending on the step size of the numerical integration and the precision of the computation, a zero derivative will be a fixed point of the numerical integration iteration.

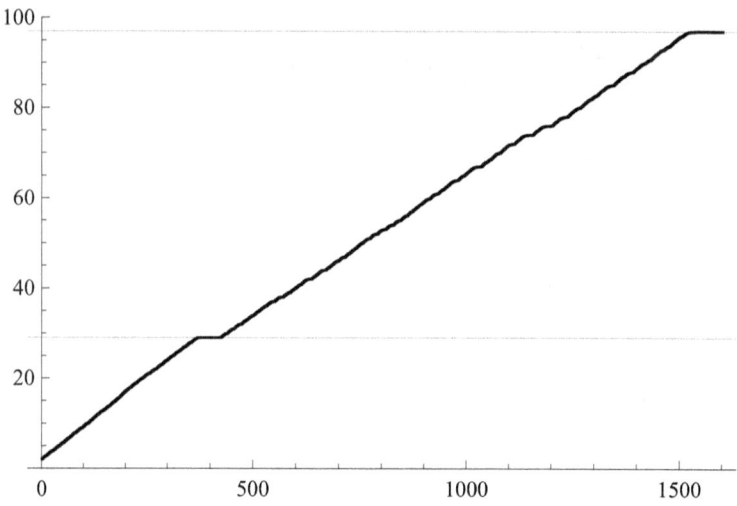

FIGURE 6.6. 1,750 Iterations of $29 \times 97$ with $h = 0.1$

For numerical integration we first consider the Euler iteration

$$x_{i+1} = x_i + hf(x_i)$$

where we take $x_0 = 2$ and $h = 0.1$. Figure 6.6 plots the values of the first $1,750$ iterates for $n = 29 \times 97$. This numerical integration of $f$ shows zero derivatives at 29 and 97 as expected.

In Figure 6.7 shows the Euler iterations for $p = 229$, $q = 541$ and $h = 0.1$ One can dimly see that the number of iterations to get from one integer to the next is not constant.

Figure 6.8 is a plot of the number of Euler iterations needed to get from each abscissa integer to the next. The ordinate value at the abscissa $i$ is the count of the number of Euler iterations needed to get from $i$ to $i + 1$; i.e. the number of Euler iterations in the interval $[i, i+1)$. For example, the cross-hairs say that it took 15 Euler iterations to get from 80 to 81.

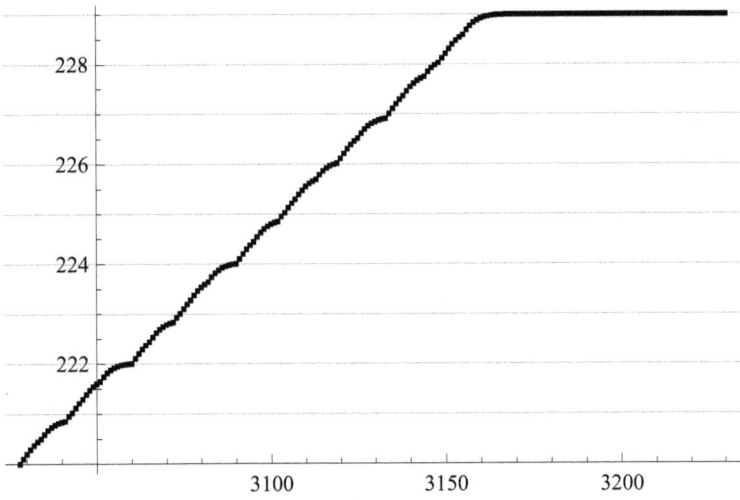

FIGURE 6.7. Euler Iterations Approach 229 for $n = 229 \times 541$

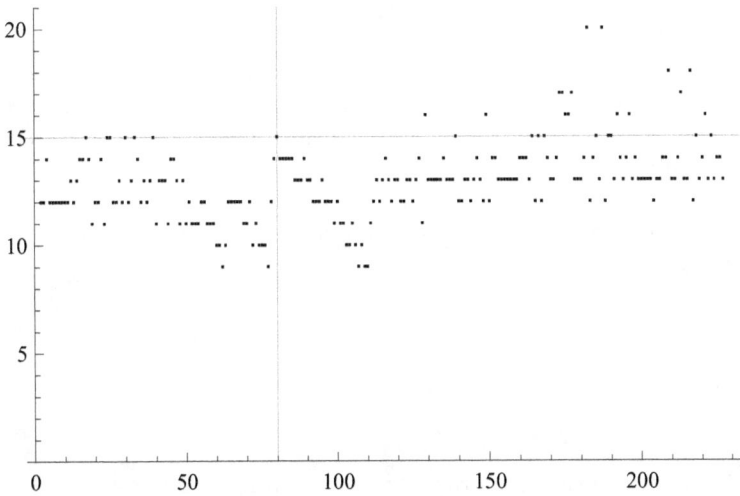

FIGURE 6.8. Euler Iterations Across Integer Intervals for $n = 229 \times 541$

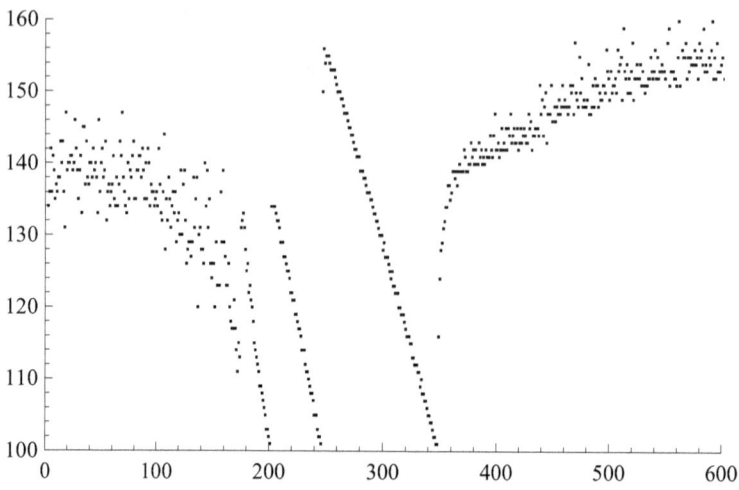

FIGURE 6.9. Euler Iterations Across Integer Intervals for $n = 1,283 \times 9,467$

Figure 6.9 shows a plot of the number of iterations to get across each integer interval in the application of the Euler algorithm to $1,283 \times 9,467$ with $h = 0.01$ as the iteration heads toward $1,283$. One can't help but being struck by the linear elements.

Figure 6.10 is the same plot as Figure 6.9 with vertical grid lines drawn at the abscissae:

$$\sqrt{\frac{hpq}{i}} \text{ for } i = 1, ..., 5$$

and horizontal grid lines drawn at the following ordinates:

$$77.9 = \frac{1.57409}{h^{0.9974}},$$

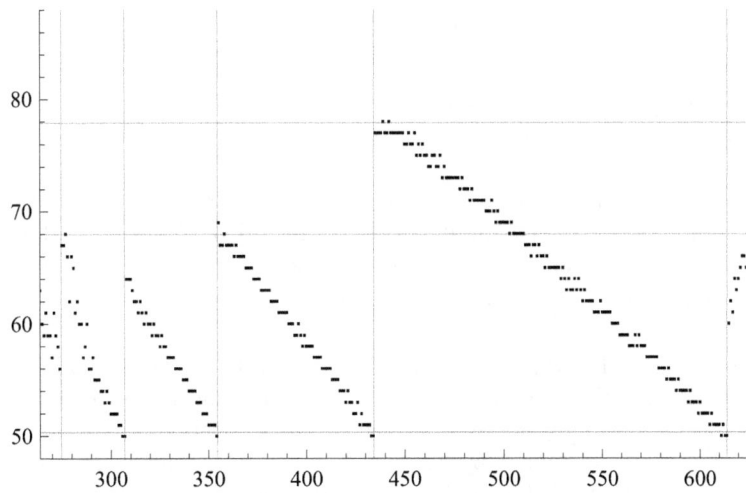

FIGURE 6.10. Euler Iterations Across Integer Intervals for $n = 1,283 \times 9,467$

$$67.9 = \frac{1.36968}{h^{0.9981}}$$

and

$$50.3 = \frac{1}{h^{1.00165}}.$$

## 6.7. Final Words

At the end of the preface to his book *The Mathematics of Great Amateurs*, Julian Coolidge says

> My friend Professor Archibald of Brown University [...] truly remarked that the number of men included could easily be doubled or trebled. I should be most happy to see someone undertake this interesting task. [**48**]

Coolidge counts John Napier, William Brouncker, Blaise Pascal, and Guillaume L'Hospital among the great amateurs. In this history of the mediant we've seen mathematicians of many levels of expertise work together in consideration of a curious property of vulgar fractions. To the person that accepts Coolidge's challenge I would suggest that Charles Haros be added to Coolidge's list of great amateurs.

# Landau's Proof of Franel's Two-Dimensional Integral

For the identity he uses in this process

$$\int_0^1 f(ax)f(ba)dx = \frac{(a,b)^2}{12ab}$$

$(f(x) = [x] - x + 1/2$, $a$ and $b$ are positive whole integers), I am communicating the following elementary proof here; it is based upon the well-known identity

$$\sum_l f\left(x + \frac{l}{m}\right)f(mx)$$

where $l$ intersects with a complete residue system $\quad \bmod m$.

1. Let $(a,b) = 1$. Then

$$\int_0^1 f(ax)f(bx)dx = \sum_{h=0}^{a-1} \int_{\frac{h}{a}}^{\frac{h+1}{a}} f(ax)f(bx)dx$$

$$= \sum_{h=0}^{a-1} \int_0^1 f\left(a\left(\frac{y}{a} + \frac{h}{a}\right)\right) f\left(b\left(\frac{y}{a} + \frac{h}{a}\right)\right) \frac{dy}{a} = \frac{1}{a}\int_0^1 f(y)f\left(\frac{by}{a} + \frac{by}{a}\right) dy$$

$$\frac{1}{a}\int_0^1 f(y) \sum_{h=0}^{a-1} f\left(\frac{by}{a} + \frac{bh}{a}\right) dy = \frac{1}{a}f(y)f(by)dy$$

$$= \frac{1}{ab}\int_0^1 f(x)f(x)dx = \frac{1}{ab}\int_0^1 \left(\frac{1}{2} - x\right)^2 dx = \frac{1}{12ab}.$$

2. Let $(a, b) = c > 1$, $a/c = \alpha$, $b/c = \beta$. Since $(\alpha, \beta) = 1$, then according to 1.

$$\int_0^1 f(ax)f(bx)dx = \int_0^1 f(\alpha \cdot cx)f(\beta \cdot cx)dx = \frac{1}{c}\int_0^c f(\alpha y)f(\beta y)dy$$

$$= \int_0^1 f(\alpha y)f(\beta y)dy = \frac{1}{12\alpha\beta} = \frac{c^2}{12ab}.$$

# "Some Consequences of the Riemann Hypothesis"

*Some consequences of the hypothesis that Riemann's $\zeta(s)$ function has no zeros in the half-plane $\Re(s) > \frac{1}{2}$. Note by J-E. LITTLEWOOD, presented by Mr. Émile Picard.*

The following theorem was independently discovered by several authors:

$$f(x) = \sum_{n=0}^{\infty} a_n(s - s_0)^n$$

*being convergent for $|s - s_0| \leq r_3|$ for$r_1 \leq r_2 \leq r_3$, we have the inequality:*

$$[M(r_2)]^{\lg \frac{r_3}{r_1}} \leq [M(r_1)]^{\lg \frac{r_3}{r_2}} \leq [M(r_1)]^{\lg \frac{r_2}{r_1}}$$

*where $M(r)$ designates the maximum of $|f(s)|$ on the circumference $|s - s_0| = r$.*

With the help of this theorem, I deduce the following from Riemann's hypothesis that $\zeta(s) \neq 0$ for $\Re(s) > \frac{1}{2}$:

$$\lg \zeta(\sigma + it) = O\left[(\lg t)^{2(1-\sigma)+\varepsilon}\right],$$

*and in the same way for*

$$\frac{1}{2} + \delta \leq \sigma \leq 1$$

*$\delta$ and $\varepsilon$ designating arbitrarily small positive values that are independent of each other.*

Let us posit that

$$s_0 = \sigma_0 + it, \qquad r_1 = \sigma_0 - (1 + \tfrac{1}{2}\delta), \qquad r_2 = \sigma_0 - \sigma_1, \qquad r_3 = \sigma_0 - \tfrac{1}{2}(1 + \delta).$$

Then according to well-known theorems with respect to $\zeta(s)$, we have the relations

$$M(r_3) = O(\lg t), M(r_1) = O(1),$$

and by applying (1) using a suitable value for $\sigma_0$ we obtain a result that is equivalent to (2).

Let us now specify in (2) that $\varepsilon = \delta$; this result in

$$|log\zeta(s)| < K(\lg t)^{1\delta}$$

and thus

$$\left|\frac{1}{\zeta(s)}\right| < \exp \lg t K(\lg t)^{-\delta} = t^{K(\lg t)^{-\delta}} = O(t^\varepsilon)$$

$\varepsilon$ being arbitrarily small. In an analogous fashion,

$$\left|\frac{1}{\zeta(s)}\right| = O(t^\varepsilon) \text{for} \sigma > \frac{1}{2} + \delta$$

According to a well-known theorem, we can now arrive at the following result: *The Dirichlet series*

$$\sum_{n=0}^{\infty} \mu(n)n^{-s} = \frac{1}{\zeta(s)}$$

*and*

$$\sum_{n=0}^{\infty} \lambda(n)n^{-s} = \frac{\zeta(2s)}{\zeta(s)}$$

*converge in the half plane* $\Re(s) > \frac{1}{2}$.

We obtain analogous results from Dirichlet series of a more general character. For example: *(5) The straight line $\sigma = \alpha$ being the line of convergence of the series $f(s) = \sum a_n n^{-s}$ either the function $f(s)$ has singularities in the entire domain $\sigma > \alpha - \delta$, or it takes each assigned value an infinite number of* *times in all of these domains, or, finally, one has in all of these domains for every value of k,*

$$\limsup |f(s)s^{-k}| = \infty$$

It remains for us to decide whether this final alternative is superfluous and in what measure the Dirichlet series may be generalized.

More subtle considerations have led me to the following theorems, of which the first provides us with a more precise version of (2).

*For*

$$\frac{1}{2} + \frac{\delta}{\lg_2 t} \leqq \sigma \leqq t$$

*we uniformly have the relation*

$$\lg \zeta(s)\ \lg \zeta(\sigma + it) = O\left[\left(\frac{\lg t \lg_2 t}{\lg_3 t}\right)^{2(1-\sigma)} \lg_2 t\right],$$

*where*

$$\lg_2 = \lg \lg t, \ldots$$

*In addition (still based upon the Riemann hypothesis)*

$$\lg \zeta(1 + it) = \lg \left[O(lg_2 t \lg_3 t)\right],$$

*such that* $\zeta(1 + it)$ *and* $\frac{1}{\zeta(1+it)}$ *are of the form*

$$O(\lg 2t \lg_3 t).$$

If, instead of Riemann's hypothesis, we only assume that the abscissas of the zeros of $\zeta(s)$ are all $< \Theta < 1$, then theorems (2)', (3) and (3)' would have to be replaced by analogous theorems, whereas (6) would remain unaltered.

Finally, by using a theorem from Mr. Bohr and Mr. Landau, according to which $|\zeta(1 + ti| > K^{-1}\lg_2 t$ for *particular* large values of $t$, I deduce the following theorem, (regardless whether Riemann's hypothesis is true or not):

*(7) Either the function $\zeta(s)$, or the function $\zeta'(s)$ has an infinity of zeros in the half plane $\sigma > 1 - \delta$, $\delta$ being an arbitrarily small positive quantity.*

# Bibliography

[1] G. L. Alexanderson, *The Random Walks of George Polya*. New York: Mathematical Association of America, 2000.

[2] Anonymous, *Métrologies Constitutionnelle et Primitive, Comparées entre elles et avec la Métrologie d'Ordonnances*. Paris: H. J. Jansen, 1801, vol. 1.

[3] ——, "Circulating Decimals. Notice of the Late Henry Goodwyn, Esq." *Mechanics' Magazine*, no. 244, pp. 197–199, 1828.

[4] R. C. Archibald, "Mathematical Tables in Phil. Mag." *Math. Tables and Other Aids to Comp.*, vol. 1, no. 5, pp. 135–141, April 1944.

[5] ——, *Mathematical Table Makers*. New York: Scripta Mathematica, 1948.

[6] ——, "Review: The Farey Series of Order 1025 by E.H. Neville," *Math. Tables and Other Aids to Comp.*, vol. 5, no. 35, pp. 133–160, 1951.

[7] V. I. Arnol'd, "On Teaching Mathematics," 1997. [Online]. Available: http://pauli.uni-muenster.de/ munsteg/arnold.html

[8] Arvind and D. E. Culler, "Dataflow Architectures," *Annual Review of Computer Science*, vol. 1, pp. 225–253, 1986.

[9] A. Aubry, "Les Logarithmes avant Neper," in *L'enseignement Mathématique*, C.-A. Laisant and H. Fehr, Eds. Paris: Gauthier-Villar, 1906, vol. 8, pp. 417–432.

[10] B. H. Baguley, "Fraction exploration device," U.S. Patent 6,840,439, January 11, 2005.

[11] P. T. Bateman, "Book Review: The Farey series of order 1025, displaying solutions of the Diophantine equation, $bx - ay = 1$," *Bull. Amer. Math. Soc.*, vol. 57, no. 4, pp. 325–326, 1951.

[12] O. Beard, "To the Editor of the MUSICAL WORLD," *The Musical World*, p. 806, December 1864.

[13] M. Bell and J. Milne, "A Working List of Mathematical Tables," in *Handbook of the Napier Tercentenary Celebration or Modern Instruments and Methods of Calculation*, E. Horsburgh, Ed. London: G. Bell and Sons, Ltd., 1914, ch. 3, pp. 47–60.

[14] R. S. Belliveau, "Lighting system incorporating programmable video feedback lighting devices and camera image rotation," U.S. Patent 6,719,433, April 13, 2004.

[15] J. Bertrand, "Mémoire sur le nombre de valeurs que peut prendre une fonction quand on y permute les lettres qu'elle renferme," *J. l'École Polytechnique*, pp. 123–140, 1845.

[16] P. J. Bickel, E. A. Hjammel, and J. W. O'Connell, "Sex Bias in Graduate Admissions: Data From Berkeley," *Science*, vol. 187, pp. 398–404, 1975.

[17] G. Blanche and E. Yowell, "A Guide to Tables on Punched Cards," *Math. Tables and Other Aids to Comp.*, vol. 5, no. 36, pp. 185–212, October 1951. [Online]. Available: http://www.jstor.org/stable/2002107

[18] C. R. Blyth, "On Simpson's Paradox and the Sure-Thing Principle," *J. Amer. Statistical Assoc.*, vol. 67, no. 338, pp. 364–366, June 1972.

[19] M. F. Bocko and Z. Ignjatovic, "System and method for image sensing and processing," U.S. Patent Application 2009/0 136 154, May 28, 2009.

[20] J. Bonnycastle, *Treatise on Algebra in Practice and Theory*. London: Nunn, 1813, vol. 2.

[21] W. Bosma, "Approximation by mediants," *Math. Comp.*, vol. 54, no. 189, pp. 421–434, 1990.

[22] C. B. Boyer, *A History of Mathematics*. New York: John Wiley & Sons, 1989.

[23] M. Bradley, *A career biography of Gaspard-Clair-Francois-Marie Riche de Prony: bridge-builder, educator and scientist*. London: Edwin Mellen Press, 1998.

[24] R. P. Brent, *Algorithms for Minimization Without Derivatives*. New York: Dover Publications, 2002.

[25] C. Brezinski, *History of Continued Fractions and Padé Approximants*. New York: Springer-Verlag, 1991.

[26] A. Brocot, "Calcul des rouages par approximation: nouvelle méthode," *Revue Chronomeétrique*, vol. 3, pp. 186–194, 1861.

[27] ——, *Berechnug der Räderübersetzungen*, 2nd ed. Leipzig: Verlag von & Korn, 1879.

[28] H. Brown and K. Mahler, "A generalization of Farey sequences: Some exploration via the computer," *J. Number Theory*, vol. 3, pp. 364–370, 1971.

[29] E. Buckingham, *Manual of Gear Design*. New York: Industrial Press, 1935.

[30] N. Burunova, *Spravochnik po matematicheskim tablitsam*, ser. Dopolnenie. Moscow: Akad. Nauk., 1959, no. 1.

[31] F. Cajori, *A History of Mathematics*, 3rd ed. London: Chelsea, 1980.

[32] C. Camus, *Cours de Mathématique*. Paris: De L'Imprimerie Royale, 1752, vol. 2.

[33] M. L. Cartwright, "Balthazar van der Pol," *J. London Math. Soc.*, vol. 35, pp. 367–376, 1919.

[34] A. L. Cauchy, "D'un Théorème Curieux sur Les Nombres," *Bulletin des Sciences, par la Société Philomatique de Paris*, vol. 3, no. 3, pp. 133–135, 1816.

[35] ——, *Cours D'Analyse de l'École Royale Polytechnique*. Paris: Debure, 1821.

[36] A. Cayley, *Report of the Committee on Mathematical Tables*. London: Taylor & Francis, 1875, pp. 305–336.

[37] W. Chang, N. Ghamrawl, and A. Swami, "System and method of efficiently representing and searching directed acyclic graph structures in databases," U.S. Patent Application 2007/0 208 693, September 6, 2007.

[38] P. Chebyshev, "Mémoire sur les nombres premiers," *Mém. Acad. Sci. St. Pétersbourg*, pp. 17–33, 1850.

[39] A. M. Chiang and S. R. Broadstone, "Portable ultrasound imaging system," U.S. Patent 5,964,709, October 12, 1999.

[40] N. Chuquet, *Le Triparty en la Science des Nombres*. Rome: Imprimerie des Sciences Mathématiques et Physiques, 1481.

[41] ——, *Problèmes numériques faisant suite et servant d'application au Triparty en la Science des nombres*. Rome: Imprimerie des Sciences Mathématiques et Physiques, 1482.

[42] A.-C. Clairaut, *Éléments d'Algèbre*. Paris: Emery, 1801.

[43] C. Cobeli and A. Zaharescu, "The Haros-Farey sequence at two hundred years," *Acta Univ. Apulensis Math. Inform.*, no. 5, pp. 1–38, 2003.

[44] M. Combes, "Image dithering based on Farey fractions," U.S. Patent Application 2009/0 066 719, March 12, 2009.

[45] L. J. Comrie, *Mathematical Tables*. London: British Astronomical Association, 1928, pp. 38–43.

[46] ——, "John Thomson, Table of Twelve-Figure Logarithms," *Math. Tables and Other Aids to Comp.*, vol. 2, no. 20, pp. 353–354, October 1947.

[47] A. Conventionale, "No. 216. Courier de L'Égalité,," in *Almanach National de France*, Paris, March 1793, p. 470.

[48] J. L. Coolidge, *The Mathematics of Great Amateurs*. Oxford: Clarendon Press, 1949.

[49] M. Croarken, "Table making by committee: British Table Makers, 1871–1965," in *The History of Mathematical Tables from Sumer to Spreadsheets*, M. Campbell-Kelly, M. Croarken, R. Flood, and E. Robson, Eds. Oxford: Oxford University Press, 2003, ch. 9, pp. 234–263.

[50] H. Davis, *Mathematical Tables*. Toronto: Encyclopedia Britannica, 1947, pp. 80–84.

[51] H. Davis and V. Fisher, *A Bibliography and Index of Mathematical Tables*. Evanston: Northwestern University, 1949.

[52] A. De Morgan, "Tables," in *The Penny Cyclopedia of the Society for the Diffusion of Useful Knowledge*, C. Knight, Ed. London: Charles Knight & Co., 1842, vol. 23, pp. 496–501.

[53] ——, "Tables," in *The Supplement to the Penny Cyclopedia of the Society for the Diffusion of Useful Knowledge*, C. Knight, Ed. London: Charles Knight & Co., 1851, vol. 2, pp. 595–605.

[54] ——, "Tables," in *Arts and Sciences or Fourth Division of "The English Cyclopaedia"*, C. Knight, Ed. London: Bradbury, Evans and Co., 1868, vol. 7, ch. Tables, pp. 976–1015.

[55] A. A. C. M. P. de Polignac, "Six propositions arithmologiques deduites du crible d'Eratosthene," *Nouvelles Annales de Mathematiques*, vol. 8, pp. 421–429, 1849.

[56] G. de Prony, "Suite des lçons d'analyse," *J. l'École Polytechnique*, vol. 4, 1796.

[57] A.-M. Décaillot, "Géométrie des tissus. Mosaïques. Échiquiers. Mathématiques curieuses et utiles," *Revue d'histoire des mathématiques*, vol. 8, pp. 145–206, 2002.

[58] R. C. F. M. Dechales, *Tractatus Proemialis de progressu Matheseos et illustribus Mathematicis*. Utrecht: Lugduni, 1690.

[59] J.-B. J. Delambre, "Analysis of the Labours of the Royal Academy of Sciences of the Institute of France during the year 1816," in *The Philosophial Magazine and Journal*, A. Tilloch, Ed. London: Richard & Arthur Taylor, 1817, vol. 49, p. 345.

[60] ——, *Histoire de L'Astronomie Moderne*. Paris: Courcier, 1821.

[61] L. Delegrive, *Manuel de Trigonométrie Pratique*. Paris: Courcier, 1806.

[62] L. E. Dickson, *History of the Theory of Numbers. Divisibility and Primality*. Washington, D.C.: Carnegie Institute, 1919, vol. 1.

[63] P. Druck, "Measurement scale for non-uniform data sampling in N dimensions," U.S. Patent 6,477,553, November 5, 2002.

[64] M. du Satory, *The Music of the Primes: Searching to Solve the Greatest Mystery in Mathematics*. New York: Harper Perennial, 2004.

[65] H. M. Edwards, *Riemann's Zeta Function*. New York: Academic Press, 1974.

[66] P. Erdős, "A Theorem of Sylvester and Schur," *J. London Math. Soc.*, vol. 9, pp. 282–288, 1934.

[67] ——, "A note on Farey series," *Quart. J. Math., Oxford Ser.*, vol. 14, pp. 82–85, 1943.

[68] J. S. Ersch, *Litteratur der Mathematik, Natur- und Gewerbs-Kunde mit Inbegriff der Kriegskunst*. Leipzig: J.H. Brodhaus, 1828.

[69] L. Euler, *Tentamen novae theoriae musicae, ex certissimis harmoniae principiis dilucide expositae*. Lausannae: Petropoli, 1739.

[70] J. Farey, *General view of the agriculture and minerals of Derbyshire; with observations on the means of their improvement drawn up for the consideration of the Board of Agriculture and Internal Improvement*. London: Sherwood, Neely & Jones, 1815.

[71] ——, "On a Curious Property of Vulgar Fractions," *The Philosophical Magazine and Journal: Comprehending the various branches of science, the liberal and fine arts, geology, agriculture, manufacturers and commerce.*, vol. 47, no. 3, pp. 385–386, 1816.

[72] ——, *A General View of the Agriculture and Minerals of Derbyshire, Volume 1. With Introduction by Trevor D. Ford and Hugh S. Torrens*. London: Peak District Mines Historical Society, 1989.

[73] J. Fauvel, R. Flood, and R. Wilson, *Music and Mathematics: From Pythagoras to Fractals*. Oxford: Oxford University Press, 2006.

[74] G. Flegg, *Numbers: Their History and Meaning*. New York: Dover Publications, 1983.

[75] G. Flegg, C. Hay, and B. Moss, *Nicolas Chuquet, Renaissance Mathematician*. Dordrecht: D. Reidel, 1985.

[76] A. Fletcher, "Review: Rectangular-Polar Conversion Tables by E.H. Neville," *Math. Gazette*, vol. 41, no. 337, p. 234, 1957.

[77] A. Fletcher, J. C. P. Miller, and L. Rosenhead, *An Index of Mathematical Tables*. London: Scientific Computing Service, 1946, vol. 1.

[78] A. Fletcher, J. C. P. Miller, L. Rosenhead, and L. J. Comrie, *An Index of Mathematical Tables (Second Edition), Part I. Index according to functions.* Oxford: Blackwell Scientific Publications, 1962, vol. 1.

[79] J. Flitcon, "A general method for solving this question," in *The Mathematical Questions, proposed in the Ladies' Diary, and their original answers, together with some new solutions, from its commencement in the year 1704 to 1816*, T. Leybourn, Ed. J. Mawman, 1817, vol. 1, pp. 399–400.

[80] T. D. Ford and H. S. Torens, "A Farey Story: the Pioneer Geologist John Farey (1766 – 1826)," *Geology Today*, vol. 17, no. 2, pp. 59–68, March–April 2001.

[81] D. H. Fowler, "An Approximation Technique, and its Use by Wallis and Taylor," *Archive for History of Exact Sciences*, vol. 41, no. 3, pp. 189–233, September 1991.

[82] ——, "An objective and practival method for describing and understanding ratios," *Mathématiques et Sciences Humaines*, vol. 124, pp. 5–18, 1993.

[83] ——, "Continued Fractions," in *Companion encyclopedia of the history and philosophy of the mathematical sciences*, I. Grattan-Guinness, Ed. London: Routledge, 1994, ch. 6.3, pp. 730–704.

[84] ——, *The Mathematics of Plato's Academy: A New Reconstruction.* Oxford: Oxford University Press, 1999.

[85] J. Franel, "Les suites de Farey et les problèmes des nombres premiers," *Göttinger Nachrichten*, pp. 198–206, 1924.

[86] A. Fujii, "On the Farey series and the Riemann hypothesis," *Comment. Math. Univ. St. Pauli*, vol. 54, no. 2, pp. 211–235, 2005.

[87] ——, "On the Farey series and the Hecke *L*-functions," *Comment. Math. Univ. St. Pauli*, vol. 56, no. 2, pp. 97–162, 2007.

[88] J.-G. Garnier, *Analyse Algébrique, faisant suire aux Élémens d'Algébre.* Paris: Courcier, 1804.

[89] ——, *Analyse Algébrique, faisant suire a la Premiére section de l'algébre.* Paris: Courcier, 1814.

[90] C. B. Gentry, "Encryption and signature schemes using message mappings to reduce the message size," U.S. Patent Application 2006/0 159 259, July 20, 2006.

[91] D. Gilles, *Bibliographies on Numerical Calculations.* London: Scientific Computing Services, 1954.

[92] J. W. L. Glaisher, *Report of the Committee on Mathematical Tables.* London: Taylor & Francis, 1873.

[93] ——, "On circulating decimals with special reference to Henry Goodwyns Table of Circles and Tabular series of decimal quotients (London 1818-1823)," *Proc. Cambridge Philos. Soc.*, vol. 30, pp. 185–206, 1878.

[94] ——, "On a property of vulgar fractions," *Phil. Mag. (Series 5)*, vol. 7, pp. 321–336, 1879.

[95] J. Glaisher, W. Bickley, C. Gwyther, J. Miller, and E. Ternouth, *Tables of powers giving integral powers of integers.* London: British Association for the Advancement of Science, 1940, vol. 9.

[96] A. Gluchoff, "Pure mathematics applied in early twentieth-century America: The case of T.H. Grönwall, consulting mathematician," *Historica Mathematica*, vol. 32, no. 3, pp. 312–357, August 2005.

[97] S. W. Golomb, "Formulas for the next prime," *Pacific J. Math.*, vol. 63, pp. 401–404, 1976.

[98] H. Goodwyn, *The brewer's assistant: containing a variety of tables, calculated to find, with precision, the value,quantity, weight, &c. of the principal articles purchased, sold or retained, in a brewing trade.* London: J. Davis, 1796.

[99] ——, "On the new Measures of France," *Journal of Natural Philosophy, Chemistry and the Arts*, vol. 4, pp. 164–165, July 1800.

[100] ——, "A Section and Description of a Machine that will raise a Body of Water to any Height, not exceeding the Height of a Column that will counterbalance the Pressure of the Atmosphere (say 30 Feet) by the Descent of Part of the same Body of Water, through a somewhat greater height, and aided by the Pressure of the Atmosphere," *Journal of Natural Philosophy, Chemistry and the Arts*, vol. 4, pp. 165–167, July 1800.

[101] ——, "Construction and Use of an universal Table of Interest," *Journal of Natural Philosophy, Chemistry and the Arts*, vol. 4, pp. 433–438, January 1801.

[102] ——, "On raising Water by the Engine," *Journal of Natural Philosophy, Chemistry and the Arts*, vol. 4, pp. 342–344, 1801.

[103] ——, "Curious properties of prime numbers, taken as the divisors of unity," *Nicholson Journal*, vol. 1, pp. 314–316, 1802.

[104] ——, *A Table to compare a new System of English with the new System of French Measures and Weights*. London: Royal Society, 1803.

[105] ——, *The First Centenary of a Series of Concise and Useful Tables of all the Complete Decimal Quotients, which can arise from dividing a unit, or any whole Number less than each Divisor by all Integers from 1 to 1024*. Private Distribution, 1816.

[106] ——, *The First Centenary of a Series of Concise and Useful Tables of all the Complete Decimal Quotients, which can arise from dividing a unit, or any whole Number less than each Divisor by all Integers from 1 to 1024 To which is now added a Tabular Series of Complete Decimal Quotients for all the Proper Vulgar Fractions of which when in their lowest terms, neither the Numerator nor the Denominator is greater than 100: with the equivalent vulgar fractions prefixed*. London: Payne & Foss, 1818.

[107] ——, *Tables of Complete decimal Quotients with their Reciprocals, corresponding Vulgar Fractions, and Logarithms, preceeded by an Index*. Private Distribution, 1820.

[108] ——, *A very concise, yet strictly accurate method, for finding the interest of any given sum : at any given rate, for any given number of days*. London: A. Constable, 1820.

[109] ——, *Introduction to a synoptical table of English and French lineal measures. A Synoptical table of ancient and modern English and French lineal measures, arranged and calculated by Henry Goodwyn*. London: J.M. Richardson, 1821.

[110] ——, *A diagonal table to facilitate the comparison of measures of capacity, which result naturally from the proposed imperial gallon*. London: J.M. Richardson, 1823.

[111] ——, *A Table of the Circles arising from the Division of a Unit, any other whole Number by all the Integers from 1 to 1024, being all the decimal Quotients that can arise from this source*. London: J.M. Richardson, 1823.

[112] ——, *A Tabular Series of Decimal Quotients for all proper Vulgar Fractions, in which when in their lowest Terms, neither the Numerator nor the Denominator is greater than 1000*. London: J.M. Richardson, 1823.

[113] H. S. Grant, "Additive Entities, an Extension of Farey Series," *National Mathematics Magazine*, vol. 14, no. 5, pp. 256–260, 1940.

[114] I. Grattan-Guinness, *Convolutions in French Mathematics, 1800-1840*. Basel: Birkhäuser Verlag, 1990, vol. 1.

[115] ——, "Work for the hairdressers: the production of de Prony's logarithmic and trigonometric tables," *Annals of the History of Computing*, vol. 12, no. 3, pp. 177–185, 1990.

[116] ——, *The computation factory: de Prony's project for making tables in the 1790's*. Oxford: Oxford University Press, 2003, ch. 4, pp. 105–122.

[117] O. G. Gregory, *A Table of Circles, from which knowing the diameters, the areas, circumference, and sides of equal squares, are found*. London: Baldwin, Cradock, and Joy, 1825, pp. 399–407.

[118] D. A. Grier, "The Rise and Fall of the Committee on Mathematical Tables and Other Aids to Computation," *IEEE Annals of the History of Computing*, pp. 38–49, April–June 2001.

[119] ——, *When Computers Were Human*. Princeton, NJ: Princeton University Press, 2005.

[120] R. Griswold, "Designing with Farey Fractions," 2002. [Online]. Available: http://www.cs.arizona.edu/patterns/weaving/webdocs

[121] R. K. Guy, *Gaps between Primes. Twin Primes.*, 3rd ed., ser. Problem Books in Mathematics. New York: Springer-Verlag, 2004, ch. A8, pp. 31–39.

[122] M. H. T. Hack, "Fast correctly-rounding floating-point conversion," U.S. Patent Application 2008/0 065 709, March 13, 2008.

[123] G. H. Hardy, *A Mathematician's Apology*. Cambridge: Cambridge University Press, 1940.

[124] G. H. Hardy and E. Wright, *An Introduction to the Theory of Numbers*. Oxford: Oxford University Press, 1979.

[125] C. Haros, *Instruction abrégée aux nouvelles mesures qui doivent être introduites dans toute la République au $1^{er}$ Vendémiaire an 10: avec Tables de rapports et de reéductions*, 1st ed. Paris: Firmin Didot, 1801.

[126] ——, *Instruction abrégée aux nouvelles mesures qui doivent être introduites dans toute la République au $1^{er}$ Vendémiaire an 10: avec Tables de rapports et de reéductions*, 3rd ed. Paris: Firmin Didot, 1801.

[127] ——, *Comptes faits à la maniére de Barême sur les nouveaux Poids et Mesures, aves les prix proportionnels à l'usage des commerçans etc.* Paris: Courcier, 1802.

[128] ——, "Tables pour évaluer une fraction ordinaire avec autant de decimales qu'on voudra; et pour trouver la fraction ordinaire la plus simple, et qui approche sensiblement d'une fraction décimal," *J. l'École Polytechnique*, vol. 6, no. 11, pp. 364–368, 1802.

[129] ——, *Tables de logarithme à l'usage dé s ingénieurs du cadastre et des éleves qui se destinent à l'École Polytechnique*. Paris: Courcier, 1806.

[130] ——, "Tables de Logarithmes et des tables pour la conversion des nouveaux poids et mesures," in *Arithmétique de Bezout*, A. Reynaud, Ed. Paris: Courcier, 1806.

[131] C. Haros, M. Plausol, and M. Bauzon, "Tables de logarithme à l'usage des ingéieurs du cadastre de des élèves qui se destinent à lÉcole Polytechnique," in *Trigonométrie Analytique, précédee de la Théorie des logarithmes Analytique*, A. Raynaud, Ed. Paris: Courcier, 1805.

[132] J. I. Hawkins, *A Treatise on the Teeth of Wheels Demonstrating the best forms which can be given to them for the purposes of machiner; such as mill-work and clock-work, and the art of finding their numbers*. Manchester: James S. Hodson, 1837.

[133] J. C. Heilbronner, *Historia Matheseos Universæ*. Lipsiæ: Joh. Friderici Gleditschii, 1742.

[134] J. Henderson, *Bibliotheca tabularum mathematicarum: Being a descriptive catalogue of mathematical tables. Part I.* Cambridge: Cambridge University Press, 1926.

[135] P. Henrici, "A Subroutine for Computations with Rational Numbers," *J. ACM*, vol. 3, no. 1, pp. 6–9, 1956.

[136] I. S. Hornsey, *A History of Beer and Brewing*. London: Royal Society of Chemistry, 2004.

[137] C. Hutton, *History of Trigonometric Tables*. London: Rivingtons, 1811, ch. 1.

[138] ——, *Tracts on Mathematical and Philosophical Subjects*. London: Rivingtons, 1812.

[139] J. Itard, "Nicolas Chuquet," in *Biographical dictionary of Mathematicians : reference biographies from the Dictionary of scientific biography*. Scribner, 1991, vol. 1, pp. 497–503.

[140] S. Ito, "Algorithms with mediant convergents and their metrical theory," *Osaka J. Math.*, vol. 26, no. 3, pp. 557–578, 1989.

[141] G. Izawa, T. Saito, and H. Torikai, "A dependent switched capacitor A/D converter for Farey series approximation," in *Proceedings of ISCAS 2000*, vol. 5. IEEE, 2000, pp. 681–684.

[142] D. P. Jablon and J. G. Landau, "Data integrity and non-repudiation system," U.S. Patent Application 2009/0 049 299, February 219, 2009.

[143] S. A. Kahn, "The bounds of the set of equivalent resistances of n equal resistors combined in series and in parallel," *arXiv.org*, pp. 1–37, 2010. [Online]. Available: http://arxiv.org/abs/1004.3346

[144] S. Kanemitsu and M. Yoshimoto, "Farey series and the Riemann hypothesis. III," *Ramanujan J.*, vol. 1, no. 4, pp. 363–378, 1997, international Symposium on Number Theory (Madras, 1996).

[145] J. Kappraff, "The Arithmetic of Nicomachus of Gersa and its Applications to Systems of Proportions," *Nexus Network Journal*, vol. 2, pp. 41–55, 2000.

[146] A. G. Kästner, *Geschichte der Mathematik*. Lipsiæ: Teubner, 1796.

[147] C. Knott, *Napier Tercentenary Memorial Volume*. London: Longmans, 1915.

[148] D. E. Knuth, *The Art of Computer Programming, Seminumerical Algorithms*, 3rd ed. New York: Addison-Wesley Longman, 1998, vol. 2.

[149] P. Kornerup and D. W. Matula, "A feasibility analysis of fixed-slash rational arithmetic," in *Proceedings of the Fourth IEEE Symposium on Computer Arithmetic*. IEEE, 1978, pp. 39–47.

[150] S. Lacroix, *Complément des Élémens d'Algébre*, 3rd ed. Paris: Courcier, 1804.

[151] ——, *D'Arithmétique a l'Usage de l'École Centrale des Quatre-Nations*, 7th ed. Paris: Courcier, 1807.

[152] J. C. Lagarias, "The Riemann hypothesis: arithmetic and geometry," in *Surveys in Noncommutative Geometry*, N. Higson and J. Roe, Eds., Clay Institute. Providence, R.I.: Amer. Math. Soc., 2007, pp. 127–141.

[153] J. L. Lagrange, "Essai d'analyse numérique sur le transformation des fractions," *J. l'École Polytechnique*, vol. 2, pp. 93–115, 1798.

[154] ——, *Traité de la résolution des equations numériques de tous les degrés*, 2nd ed. Paris: Gauthier-Villar, 1808.

[155] J. d. Lalande, *Bibliographie Astronomique, avec l'histoire de l'Astronomie*. Paris: Imprimerie de la République, 1803.

[156] S. Lamassé, "Une Utilisation Précoce de l'Algébre en France au XV$^e$ Siécle. Note sur le Manuscrit 1339 de la Bibliothéque Nationale de France," *Revue d'Historie des Mathématiques*, no. 11, pp. 239–255, 2005.

[157] E. G. H. Landau, "Bemerkungen zu der vorstehenden Abhandlung von Herrn Franel," *Göttinger Nachrichten*, pp. 202–206, 1924.

[158] W. Langford, T. Broadbent, and R. Goodstein, "Obituary: Professor Eric Harold Neville, M.A., B.Sc." *Math. Gazette*, vol. 48, no. 364, pp. 131–145, May 1964.

[159] R. Laubenbacher, G. McGrath, and D. Pengelley, "Lagrange and the solution of numerical equations," *Historia Mathematica*, vol. 28, pp. 220–231, 2001.

[160] T. Lavernède, "Analyse indéterminée. Recherche systématique des formules les plus propres à calculer les logarithmes," *Annales de Mathématiques pures et appliquées*, vol. 1, pp. 78–100, 1810.

[161] A. Lebedev and R. Fedorova, *A Guide to Mathematical Tables*. Elmsford, NY: Pergamon Press, 1960.

[162] C. Lee and F. Roberts, "A Comparision of Algorithms for Rational $l_\infty$ Approximaton," *Math. Comp.*, vol. 27, no. 121, pp. 111–121, 1973.

[163] M. Lefort, "Description des Grandes Tables Logarithmiques et Trigonométriques, Calculées au Bureau du Cadastre, sous la Direction de Prony, et Exposition des Méthodes et Procédés mis en Usage pour leur Construction." *Annales de l'Observatoire de Paris*, vol. 4, pp. 123–150, 1858.

[164] D. H. Lehmer, *Guide to the tables in the theory of numbers*. Washington, D.C.: National Research Council, 1941.

[165] H. Lennerstad and L. Lundberg, "Generalizations of the floor and ceiling functions using the Stern-Brocot tree," Blekinge Institute of Technology, Tech. Rep. 2006:02, 2006.

[166] J. E. Littlewood, "Quelques conséquences de lhypothèse que la fonction $\zeta(s)$ de Riemann na pas de zéros dans le demi-plan $\Re(s) > \frac{1}{2}$," *C.R. Acad. Paris*, vol. 154, pp. 263–266, 1912.

[167] É. Lucas, *Théorie des Nombres*. Paris: Gauthier-Villar, 1891.

[168] S. Lui, "An Interview with Vladimir Arnol'd," *Notices of the AMS*, vol. 44, no. 4, pp. 432–438, April 1997.

[169] R. D. Lysons, *The Environs of London*. London: T. Cadell, Jun. and W. Davies, 1796, vol. 4.

[170] H. Made, "Über Fareysche Doppelreihen," Ph.D. dissertation, Hessischen Ludwigsuniversität zu Giessen, Giessen, 1903.

[171] R. Mansuy, "Les calculs du citoyen Haros," Lycée Louis le Grand, Paris, Tech. Rep., May 2008. [Online]. Available: http://www.dma.ens.fr/culturemath/

[172] P. Mathias, *The Brewing Industry in England, 1700-1830*. Cambridge: Cambridge University Press, 1959.

[173] D. W. Matula and P. Kornerup, "A feasibility analysis of binary fixed-slash and floating-slash number systems," in *Proceedings of the Fourth IEEE Symposium on Computer Arithmetic*. IEEE, 1978, pp. 29–38.

[174] ——, "Finite precision rational arithmetic: slash number systems," *IEEE Trans. on Computers*, vol. C-34, no. 1, pp. 3–18, 1985.

[175] J. May, "Question 281," *Ladies Diary*, p. 34, 1747.

[176] A. E. Mayer, "On neighbours of higher degree in Farey series," *Quart. J. Math., Oxford Ser.*, vol. 13, pp. 185–192, 1942.

[177] R. Mehmke and M. d'Ocagne, *Calculs Numériques*. Berlin: Wilhelm Franz Meyer, 1909, no. 4, pp. 196–452.

[178] R. Mehmke, *Numerisches Rechnen*. Berlin: Wilhelm Franz Meyer, 1902, no. 2, pp. 938–1079.

[179] C. W. Merrifield, "Mathematical Questions with their Solutions," *Math. Quest. Educat. Times*, vol. 91, pp. 92–95, 1868.

[180] H. E. Merritt, *Gear Trains including a Brocot Table of Decimal Equivalents and a Table of Factors of all useful Numbers up to 200,000.* London: Pitman, 1947.

[181] A. F. Möbius, *Der barycentrische calcul.* Berlin: Johann Ambrosius Barth, 1827.

[182] D. Moser, "The star-chromatic number of planar graphs," *J. Graph Theory*, vol. 24, no. 1, pp. 33–43, 1997.

[183] F. G. A. Murhard, *Bibliotheca Mathematica.* Lipsiæ: Teubner, 1797.

[184] H. Nakada and R. Natsui, "Some metric properties of α-continued fractions," *Journal of Number Theory*, vol. 97, no. 2, pp. 287–300, December 2002.

[185] E. H. Neville, "The structure of Farey series," *Proc. London Math. Soc. (2)*, vol. 51, pp. 132–144, 1949.

[186] ——, *The Farey Series of Order 1025, Displaying Solutions of the Diophantine Equation bx − ay = 1*, ser. Royal Society Mathematical Tables. Cambridge: Cambridge University Press, 1950, vol. 1.

[187] ——, *Rectangular-Polar Conversion Tables.* London: Royal Society, 1956, vol. 20.

[188] N. Nielsen, *Géomètres Français du Dix-Huitième Siècle.* Copenhagen: Levin & Munksgaard, 1935.

[189] M. d. Ocagne, "Sur certaines suites de fractions irréductibles," *Annales de la Société scientifique de Bruxelles*, pp. 90–108, 1885.

[190] J. Ozanam, J. Audierne, and C. Haros, *Traité de l'arpentage et du toisé, ou, Méthode courte et facile pour arpenter et partager toutes sortes de terreins, et toiser toutes sortes d'étendues.* Paris: Firmin Didot, 1803.

[191] R. M. Page, *14000 gear ratios; tabulated ratios presented in common fractional and decimal forms and in differently arranged sections to facilitate the solution of all classes of gear-ratio problems.* New York: Industrial Press, 1942.

[192] K. Pearson, A. Lee, and L. Bramley-Moore, "Contributions to the Theory of Evolution. VI. Genetic (Reproductive) Selection: Inheritance of Fertility in Man, and of Fecundity in Thoroughbred Racehorses," *Phil. Trans. A*, vol. 192, pp. 257–330, 1899.

[193] R. C. Penner, "Methods of digital filtering and multi-dimensional data compression using the Farey quadrature and arithmetic, fan, and modular wavelets," U.S. Patent 7,158,569, January 2, 2007.

[194] J. Peters, A. Lodge, E. Ternouth, E. Gifford, and L. J. Comrie, *Cycles of reduced ideals in quadratic fields.* London: British Association for the Advancement of Science, 1935, vol. 4.

[195] J. Peters, *Brocot Tabelle.* Berlin: Ernst & Korn, 1922, ch. 2, pp. 24–61.

[196] J. L. Petit, "Géneralisation du paradoxe de Simpson," *Revue de statistique appliquée*, vol. 40, no. 3, pp. 47–61, 1992.

[197] A. Picton, "Les ingéineurs et la mathématisation. L'example du génie civil et de la construction," *Revue d'historie des sciences*, vol. 42, no. 1, pp. 155–172, 1989.

[198] C. Pisot, "Suites de Farey," in *Séminaire Delange-Pisot-Poitou. Théorie des nombres.* Numdam, 1960, vol. 2, pp. 1–7.

[199] J. C. Poggendorff, *Biographisch-literarisches Handwörterbuch zur Geschichte der exacten Wissenschaften.* Leipzig: Verlag von Johann Ambros Barth, 1863.

[200] H. Polachek, "History of the Journal Mathematical Tables and other Aids to Computation, 1959-1965," *IEEE Annals History of Computing*, vol. 17, no. 3, pp. 67–74, 1995.

[201] F. D. Powell, "Method and apparatus for sampling broad band spectra in fuel quantity measurement systems," U.S. Patent 4,542,336, September 17, 1985.

[202] R. Rado, "Review: The Farey series of order 1025 by E.H.Neville," *Math. Gazette*, vol. 36, no. 315, pp. 60–61, 1952.

[203] S. Ramanujan, "A Proof of Bertrand's Postulate," *J. Indian Math. Soc.*, vol. 110, pp. 181–182, 1919.

[204] A. L. Ramel, *Système Métrique, où Instruction Abrégée sur les Nouvelles Mesures*. Paris: Girardet, 1808.

[205] R. A. Rasch, "Farey Systems of Musical Notation," *Listen 2*, pp. 31–41, 1988.

[206] G. Révész, *Introduction to the Psychology of Music*. Norman, OK: University of Oklahoma Press, 1954.

[207] A. Reynaud, *Traité d'Arithmétique a L'Usage des Ingéineures du Cadastre*. Paris: Courcier, 1804.

[208] D. D. Reynolds and R. A. Slizynski, "Coherent sampling digitizer system," U.S. Patent 5,708,432, January 13, 1998.

[209] F. Rigg, "Recent advances in mathematical statistics; bibliography of mathematical statistics, 1940-42," *J. Roy. Stat. Soc. Ser. A*, vol. 109, pp. 395–450, 1946.

[210] J. Roggio, *Bibliotheca Mathematica sive Critiçus Librorun Mathematicorum*. Turingæ, 1830.

[211] F. A. Ryde, *Aspects of the greatest integer function – Part I*. Stockholm: Almquist & Wiksell, 1973.

[212] E. Sang, "Fractions, Tabulation of all, between Given Limits," *Tran. R. Soc. Edinb.*, vol. 28, p. 287, 1879.

[213] ——, "Remarks on the Great Logarithmic and Trigonometrical Tables computed in the Bureau du Cadastre under the direction of M. Prony," *Proceedings of the Royal Society of Edinburgh*, pp. 421–436, November 1872 to July 1875.

[214] S. S. Schurr, "Method of operating an integrated circuit tester employing a float-to-ratio conversion with denominator limiting," U.S. Patent Application 2007/0 115 734, May 24, 2007.

[215] K. Schütte, *Index Mathematischer Tafelwerke und Tabellen aus allen Gebieten der Naturwissenschaften*. Munich: R. Oldenbourg, 1955.

[216] M. Scott, "Fast Rounding in Multiprecision Floating-Slash Arithmetic," *IEEE Trans. on Computers*, vol. 38, no. 7, pp. 1049–1052, July 1989.

[217] J. Sesiano, "On an algorithm for the approximation of surds from a Provençal treatise," in *Mathematics from Manuscript to Print 1300-1600*, C. Hay, Ed. Oxford Science Publication, 1988, ch. 2, pp. 30–55.

[218] E. H. Simpson, "The Interpretation of Interaction in Contingency Tables," *Journal of the Royal Statistical Society. Series B (Methodological)*, vol. 13, no. 2, pp. 238–241, 1951.

[219] R. Šleževičienė-Steuding and J. Steuding, "Simpson's paradox in the Farey sequence," *Integers*, vol. 6, pp. A4, 9 pp. (electronic), 2006.

[220] H. J. S. Smith, "On the history of the researches of mathematicians on the subject of the series of prime numbers," *Proceedings of the Ashmolean Society*, vol. 3, no. 35, pp. 128–131, 1857.

[221] M. Spiesser, "L'Algébre de Nicolas Chuquet dans le Contexte Francais de l'Arithmétique Commerciale," *Revue d'Historie des Mathématiques*, vol. 12, pp. 7–33, 2006.

[222] E. Stanhope, "Letter relative to Dr. Callcott's Pamphlet on the Stanhope Temperment," in *The Philosophial Magazine and Journal*, A. Tilloch, Ed. Richard and Arthur Taylor, 1808, vol. 30, ch. n/a, pp. 34–36.

[223] S. B. Stechkin, "Farey sequences," *Mat. Zametki*, vol. 61, no. 1, pp. 91–113, 1997.

[224] M. A. Stern, "Über eine zahlentheoretische Funktion," *J. Reine Angew. Math.*, vol. 55, pp. 193–220, 1858.

[225] S. M. Stigler, Ed., *American Contributions to Mathematical Statistics in the Nineteenth Century*. New York: Arno Press, 1980.

[226] E. Taylor, *The Mathematical Practitioners of Hanoverian England, 1714-1840*. Cambridge: Cambridge University Press, 1966.

[227] A. Thompson, "BAASMTC now RSMTC: Final Report of Committee on Calculation of Mathematical Tables," *Math. Tables and Other Aids to Comp.*, vol. 3, no. 25, pp. 333–340, January 1949.

[228] J. Todd, "A Problem on Arc Tangent Relations," *Amer. Math. Monthly*, vol. 56, no. 8, pp. 517–528, 1949.

[229] ——, *Table of Arctangents of Rational Numbers*, ser. Applied Mathematics. Washington: National Bureau of Standards, 1951, no. 11.

[230] D. W. Tufts, "Signal processing apparatus and method for iteratively determining Arithmetic Fourier Transform," U.S. Patent 5,253,192, October 12, 1993.

[231] B. van der Pol, "An electro-mechanical investigation of the Riemann zeta function in the critical strip," *Bull. Amer. Math. Soc.*, vol. 53, no. 10, pp. 976–981, 1947.

[232] G. A. V. van Oijen, *Theorie der Algebra*. Schoonhoven: S.E. Van Nooten, 1868.

[233] J. Von Neumann, "Various techniques used in connection with random digits," Applied Mathematics Series, New York, 1951.

[234] J. Wallis, *Algebra Tractatus; Historicus & Practicus*. Oxford: Oxoniæ, 1693.

[235] R. Wallis and P. Wallis, *Index of British Mathematicians*. London: Jasprint Ltd., 1993, no. 3.

[236] A. Walther and H. Unger, *Mathematische Zahlentafeln, numerische Untersuchung spezieller Funktionen*. Office of Military Government for Germany, Field Information Agencies Technical, 1948, pp. 167–183.

[237] W. A. Webb, "The Farey series of polynomials over a finite field," *Elem. Math.*, vol. 41, no. 1, pp. 6–11, 1986.

[238] S. Winchester, *The Map That Changed the World: William Smith and the Birth of Modern Geology*. New York: Harper Perennial, 2009.

[239] S. Wolfram, *A New Science*. Champaign, IL: Wolfram Media, 2002.

[240] ——, "Method and System for Generating Signaling Tone Sequences," U.S. Patent 7,560,636, July 14, 2009.

[241] J. W. Wrench, "Review: Rectangular-Polar Conversion Tables by E.H.Neville," *Math. Tables and Other Aids to Comp.*, vol. 11, no. 57, pp. 22–51, 1957.

[242] W. F. Wucherer, *Beyträge zum allgemeinern Gebrauch der Decimalbrüche.* Carlsruhe, 1796.

[243] M. Yoshimoto, "Farey series and the Riemann hypothesis," *Sūrikaisekikenkyūsho Kōkyūroku,* no. 1060, pp. 41–46, 1998, number theory and its applications (Japanese) (Kyoto, 1997).

[244] ——, "Farey series and the Riemann hypothesis. II," *Acta Math. Hungar.,* vol. 78, no. 4, pp. 287–304, 1998.

[245] ——, "Farey series and the Riemann hypothesis. IV," *Acta Math. Hungar.,* vol. 87, no. 1-2, pp. 109–119, 2000.

[246] ——, "Abelian theorems, Farey series and the Riemann hypothesis," *Ramanujan J.,* vol. 8, no. 2, pp. 131–145, 2004.

[247] G. U. Yule, "Notes on the Theory of Association of Attributes in Statistics," *Biometrika,* vol. 2, no. 2, pp. 121–134, February 1903.

# Index

Veblen, Oswald, 123
Visicalc, 58

Wallis, John, 1, 16, 22
Wilkes, Maurice, 130
Wolfram, Steven, 112

Yule, George, 32